Introduction to
ARACHNOLOGY

Introduction to
ARACHNOLOGY

Theodore Savory
M.A., F.Z.S.

FREDERICK MULLER

First published in Great Britain 1974
by Frederick Muller Limited, Edgware Road,
London NW2 6LE

Copyright © 1974 Theodore Savory

ISBN   0 584 10144 9

Printed and Bound in Great Britain by
A. Wheaton & Co., Exeter

Dedicated in friendship to
John Sankey of Juniper Hall

These common creatures, known to all,
  How do they live, and feed, and mate?
What are their names? What does one call
These common creatures, known to all,
That run, and leap, and spin, and crawl?
  For one may well investigate
These common creatures, known to all,
  How they do live, and feed, and mate.

Evelyn Lambart

Contents

List of Illustrations

List of Illustrations

Preface

One of the interesting features of zoology during the second half of the twentieth century has been the rapidly growing attention given to the Arachnida. This has been conspicuous, not only in Europe and America, where the class has long attracted a certain number of enthusiasts, but also elsewhere, as in Japan, Russia, New Zealand, Brazil and South Africa.

Proof of this progress, if proof be sought, is to be found in the steadily rising number of published papers listed annually in the *Zoological Record*, in the formation of the Arachnological Society of South Eastern Asia (1928), in the success of the British Arachnological Society (1958), in the formation of the Centre Internationale de Documentation Arachnologique (1961) and, not without significance, in the appearance of the word 'Arachnida' in the zoology syllabuses of the chief Examination Boards.

These facts justify the production of this book. It is addressed essentially to two groups of readers; the sincere and competent naturalists looking for a group of small animals suited for special study; and also to any who, anxious to embark on a piece of research work, may find in Arachnida an opportunity for serious investigation.

Attention may be directed to the fact that this is not essentially a book about the Arachnida. The animals themselves are referred to in general terms, species and genera are not diagnosed, families are mentioned without much detail. My subject is the formation and prosecution of the science, and my intention is to guide naturalists into the ways and habits of arachnologists.

The study of arachnology, as outlined in the pages that follow, becomes an entrancing occupation as well as an unending source of self-entertainment. The individual arachnologist may hope that he is advancing zoological knowledge, but even if he does this to no more than an infinitesimally small extent, he has found something far superior to theatre-going, TV-watching or novel-reading as an escape from a life no less generous with its problems than is arachnology itself.

T.H.S.

I

On Arachnology

Arachnology, derived from the Greek *arachne*, a spider, and *logos*, a discourse, literally means a dissertation or treatise on spiders. It has become extended to stand for the study of spiders together with their nearest relations, such as harvestmen, scorpions, mites and others. It was born in 1802, when J. B. Lamarck, the well-known French biologist, separated the spiders and scorpions from the *Insecta Aptera* of Carl Linnaeus, and 'Les Arachnides', as Lamarck called them, parted company from the insects.

In rather more than a century and a half arachnology has developed into a comprehensive study of the eight-legged terrestrial small animals, and is now universally recognized as having attained the status of a science in its own right. Like all other sciences that are similarly concerned with a single group of animals, it may conveniently be divided into at least ten sections:—

1. *Anatomy*. This is the practical investigation of the study of the structure of the bodies of the creatures concerned, both outwardly, by the closest scrutiny, and inwardly, by dissection, so that the characteristics of the class may be determined.

2. *Morphology* discusses the facts so revealed, compares the organs of one order with the corresponding organs of another, and is, in fact, the theoretical side of the study of structure. These two branches are of great importance since

the specific identity of every arachnid depends entirely on the details of its outward features.

3. *Physiology* deals with the vital processes of the body, or the way in which the organs perform their necessary functions.

4. *Histology* describes the microscopic structure of the parts of the body, *i.e.* its cells and tissues.

5. *Embryology*, or the study of development from the egg, is of special interest since it reveals fugitive traces of parts not present in the creature when hatched, such as temporary limbs on the abdomen, and, in the case of spiders, includes a remarkable inversion of the curvature of the developing embryo.

6. *Ethology*, the description and interpretation of the behaviour of the living arachnid, its reflexes, instincts and so on, is of special importance for two reasons. Firstly because instinctive behaviour is very highly developed in arachnids, and secondly because there is reason to believe that fundamental differences exist between the functioning of the solid, central nervous system of an invertebrate and the hollow, dorsal system of a vertebrate. Critical analyses of the behaviour of Arachnida may well illuminate this problem.

7. *Distribution*. This includes the study of the geography of the class and its orders in order to ascertain which orders and families occur in each country and in each county therein.

8. *Ecology* deals with the relation between the animal and all the features of its surroundings, including, therefore, knowledge of the arachnids that are the occupants of each of the many different types of habitat. It is at present one of the most popular forms of outdoor arachnology.

9. *Taxonomy*. This includes the accurate description of all species, the determination of the correct name by which each shall be known, and their arrangement in a logical system of classification.

10. *Phylogeny* interprets the data of taxonomy, attempting to elucidate the probable course of evolution in the class and

in each order. There is in this a greater scope for deduction, imagination and personal opinion than anywhere else in zoology.

Since to all this the historical and palaeontological aspects may be added, it is manifest that there is sufficient scope to occupy any one arachnologist for the whole of his working life; with the added attraction that, though knowledge is increasing, so much remains to be discovered that every enthusiast is almost assured of some rewarding success.

*　　*　　*

Such, in brief outline, is the nature of arachnology. Although Lamarck was in no sense an arachnologist, there can be little doubt that he was responsible for the fact that arachnology began to take shape in France, as a result of the work of Latreille and Walckenaer.

P. A. Latreille (1762–1786) was the first zoologist in the world to hold a professorship of entomology, and his publications included the first attempt to compose a classification of spiders. C. A. Walckenaer (1771–1857) who succeeded him, was the author of a four-volume work on 'apterous insects', in which he named many of the genera still in use.

A little later the science began to grow in Germany where C. L. Koch (1778–1857) and C. W. Hahn (1786–1856) produced a sixteen-volume work, *Die Arachniden*. This may be said to mark a stage in the history of arachnology since it was the first publication to bear a title that included the word 'Arachnid' in any form. It contained over five hundred beautifully coloured plates.

In Britain arachnology was established by R. H. Meade (1814–1900), John Blackwall (1790–1881) and Octavius Pickard-Cambridge (1828-1917). Blackwall's two Ray Society volumes and Pickard-Cambridge's *Spiders of Dorset* were for many years the classics of British arachnology.

In America the first interest in the arachnids appeared south of the equator, but the first systematic work was that of N. M. Hentz (1797–1850) who was followed by J. H. Emerton (1847–1931) and

H. J. McCook (1837–1911), the author of the three-volume
American Spiders and their Spinningwork.

From the middle of the last century arachnology has spread
over Europe and America and has developed, sporadically and
somewhat slowly, in Africa, Asia and Australia. It has been characterized by a perhaps disproportionate attention to spiders, a
feature that is readily understandable. Spiders were, and are,
more conspicuous, more numerous and superficially more interesting than the members of the other orders and are, therefore, the
more likely to attract and receive investigation. This was em-
phasized more and more as travel became easier; expeditions bent
on collecting anything and everything that was to be found in the
remote corners of the world brought home an ever-growing
number of specimens at a rate greater than systematists could
hope to examine exhaustively.

The position was made no easier by the comparative lack of
communication and collaboration between the specialists in different countries, and by the independence of their work, which
resulted in a great deal of overlapping and superfluous effort.
Further, spiders and scorpions had no economic importance:
they were neither pests nor parasites to more than a very small
extent. They were therefore confined to the care of the purely
scientific zoologists, and had to compete for attention with the
very large number of other animals that called for description
and study.

Like entomology, its most formidable competitor, arachnology
is essentially a study of small animals, by which adjective it is
implied that a species that grows to a length of two centimetres
reaches a size beyond that of most of its fellows. This fact alone
repels some zoologists as much as it attracts others. Small size has
certain disadvantages for the animals, but it has practical advant-
ages for the student. A collection of a hundred birds or a hundred
mammals may well demand a special room for its accommodation,
while a collection of a thousand spiders needs little more than a
small box. Quite distinct from this, however, is the undeniable
charm that lies in the examination of the tiny; the revelations, at
first of the single lens and later of the microscope, are a constant

source of wonder and surprise, respects in which the Arachnida are supreme. These emotions in their turn generate a wish to acquire some skill in the handling of small bodies and ambition to perfect the technique that is a necessity in the arachnological laboratory.

Nor, finally, is it unreasonable to add the sense of pleasurable satisfaction that accompanies the knowledge that one's chosen field of research is slightly unusual, is different from the bird-watching, or the insect-hunting, or the dozens of other sides of natural history that occupy the multitude. There is a degree of originality about being an arachnologist, about being detectably superior to those who cannot distinguish a Pholcus from a Phalangium.

*   *   *

Much of the above, it might be objected, could be claimed by zoologists who, with equal enthusiasm, have given their time to sponges, sea-anemones, roundworms or woodlice. If the truth of this be granted—and there is no scale by which to measure the devotion of the nematologist or to compare it with that of the isopodologist—it calls for an estimate of the position of the arachnologist in the parent science of Zoology, or of the prestige of the arachnologist among his fellows.

The situation is not easily appraised. Almost every zoologist today leans towards one class or order of animals, and each could uphold its claim to importance and justify his wisdom in having chosen it. The truth is that to a dedicated zoologist every animal is interesting and all are in some way important: the penguin is not worth more study than the kingfisher, it is only less accessible to the majority. The evanescent publicity given a few years ago to the Coelacanth did not make that fish essentially more interesting than Archaeopteryx; no animal is more universally disliked than the louse, but the biology of lice makes an incomparably fascinating story.

Thus it is a fact that when two zoologists meet for the first time, an early question of either is likely to be 'What is your particular group?' There are so many kinds of animals, and they differ so widely from one another, that some limitation of one's range is certainly reasonable. To be a 'general zoologist' is to be content with the standard of a pass degree, to confess oneself a dilettante, skimming lightly over the surface of a vast tract of knowledge, with little advantage either to oneself or to others. An arachnologist should aim higher and justify a claim that arachnology gleams as brightly as any other facet in the zoological diamond, and stating his claim should admit no denial.

* * *

There are problems in arachnology that have occupied the minds of arachnologists for many years past. Some of these are now nearer solution than others, and are liable to recur at any moment, as a result of some new discovery or some change of opinion. Others are as likely to be perennial.

One of the aspects of the life of any arachnid is its vital timetable or life-cycle. A number of species of spiders and a smaller number of species from the other orders have been subjects of a special study, so that records exist of the normal time of appearance of the newly-hatched young, of the number of times it moults before becoming mature, and of the length of its life thereafter. There are some that produce one cocoon of eggs a year, others that produce two or more. Some species live for longer than a year, others do not. There are many arachnids, even among the commoner British species, for which the precise details of the life-cycle have not yet been determined.

There are straightforward matters of structure, which may be studied with greater and greater intensity, until they reach such details as the number of teeth on each side of the chelicerae, and the exact position of each spine on each joint of the four legs. It should not be thought that this is an exaggeration; the determination of a sub-family or a genus may depend in part on such points.

There are problems of function. An organ may be found, either inside the body or plainly visible on the surface, but the use to which it is put and the part it plays in the life of the animal are not always obvious, and even if the function is clear, details may always be added. A spider may have three pairs of spinnerets: all

6

produce silk, but are the threads from all pairs used in the same circumstances, and if different, to what different uses are they put?

This leads to questions arising from the various forms of behaviour. Arachnida have constantly shown themselves to be unusually good subjects for the investigation of instinct. The instincts of animals are not only of great interest in themselves, as the long history of their study testifies, but the instinctive actions of arachnids, as also of insects and crustaceans, seem strongly to suggest that there are detectable characteristics in invertebrate behaviour. These may be summed up in the statement that among invertebrates the body takes care of the mind, while in vertebrates the mind takes care of the body. Arachnida may well throw light on this antithesis.

Finally, there are constantly recurring changes in our systems of classification, in the naming of species, and in the detection of the evolutionary tracks of the orders and families. These are the theoretical problems, to be tackled in the study.

In this book consideration has chiefly been given to the needs of the British reader. Certainly, many of the arachnids from warmer lands are more remarkable in form and more dramatic in behaviour, but there are, I think, no aspects of arachnology that are not represented in this country.

2

The Class Arachnida

The animals that comprise the class Arachnida share the following characteristics:—

i. They have bodies founded on eighteen segments or somites, often protected by tergites above and sternites below, connected by softer pleural membrane.

ii. Of these somites, six form the cephalothorax or prosoma and twelve the abdomen or opisthosoma. The two parts, which contrast with the three, head, thorax and abdomen, of an insect, may be united across their whole breadth or may be joined by a narrow waist or pedicel.

iii. The cephalothorax carries six pairs of limbs or appendages. The chelicerae are the only ones in front of the mouth; they are followed by a pair of pedipalpi and four pairs of legs. Usually there are no appendages on the abdomen, but spiders have abdominal spinnerets.

iv. The reproductive organs are contained in the abdomen and open ventrally on the second abdominal somite.

v. Respiration is effected by lung-books or tracheae or both.

With the exception of some of the mites, the water spider *Argyroneta aquatica*, a few others that normally inhabit marshes and some that are semi-marine, arachnids live on dry land. Most of them tend to lie hidden during the day and to become active only at night, when they seek food or mates. They may be briefly described by saying that they are terrestrial, nocturnal, carnivorous, cryptozoic arthropods.

The class is divided into eleven living and five extinct orders. The latter are not considered in this book, in which four of the living orders are chosen for description and discussion. Since, as already mentioned, classification is partly a matter of personal opinion, several different ways of arranging the orders have been suggested. The system favoured by the present writer regards the class as having descended from several distinct ancestors: i.e. it is polyphyletic and may reasonably be divided into four sub-classes. Three of these are listed below, with the names of the four favoured orders printed in capitals.

Class ARACHNIDA

Sub-class Scorpionomorpha

Order SCORPIONES
Order PSEUDOSCORPIONES (or Chelonethida)
Order Solifugae (or Solpugida)

Sub-class Arachnomorpha

Order ARANEAE (or Araneida), the spiders
Order Amblypygi (or Phrynides)
Order Uropygi (or Thelyphonida)
Order Schizomida (or Tartarides)
Order Palpigradi (or Microthelyphonida)

Sub-class Opilionidea

Order OPILIONES (or Phalangida), the Harvestmen
Order Acari—the mites and ticks
Order Ricinulei (or Podogona)

This table calls for comment on two features, the first of which is the inclusion of alternative names for most of them, names which are often preferred by different writers. The reason for this is that there is no accepted ruling from the compilers of the International Code of Zoological Nomenclature, by which the name of an order must be established—and, indeed, the eleven orders listed above have in the past received between them no fewer than

9

seventy-five different names. The names chosen here are those recommended in a paper which I published in 1972, suggesting that (i) original Latin names should be retained where they exist (ii) the practice of the most eminent arachnologists should usually be followed (iii) a name should be descriptive of some characteristic of the order (iv) a name should be correct Latin.

Further, it is of interest, as illustrating the ways in which different writers have arranged the same orders, to notice that in 1945 Petrunkevitch placed them in two sub-classes, distinguished by the presence or absence of a thin waist or pedicel. Thus the Latigastra, or broad-waisted orders, included the scorpions, false-scorpions, harvestmen and mites; and the Caulogastra, which had thin waists, contained the other seven orders. This may be compared with the system of Dubinin who, in 1957, separated the orders among four different classes, one of which contained the scorpions, false-scorpions and four other orders. A second order held the harvestmen and spiders in three orders, whilst the mites, in three orders, occupied a class by themselves.

To the five diagnostic features briefly given at the beginning of this chapter, the following descriptions may be added.

The scorpions are the only arachnids that produce living young; the females of all the other orders lay eggs varying in number from one to over a thousand at a time. Sometimes the eggs are laid underground, sometimes in the shelter of a stone slab or under the bark of a tree and, among spiders and others, often enclosed in a cocoon, more or less insulated with a layer of wadding. For the short-lived species, the eggs may be the only form in which the species exists during the winter.

From the eggs the young hatch as nymphs, that is to say, they resemble their parents in all essentials except size and the ability to reproduce. They are not larvae, differing in appearance and destined to undergo the complete change of form that is known as metamorphosis.

Their growth is effected by periodic castings of the skin, known as moultings or ecdyses, often a serious and laborious undertaking which is preceded by a period of starvation. In the process the hard and tightening exoskeleton splits round the edges, the

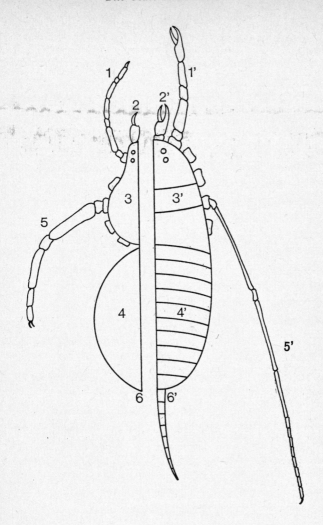

Fig. 1. Alternatives in arachnid anatomy.

1. Leg-like pedipalp. 1'. Chelate pedipalp. 2. Chelicera of two parts. 2'. Chelicera of three parts. 3. Uniform carapace. 3'. Segmented carapace. 4. Uniform abdomen. 4'. Segmented abdomen. 5. Short leg of seven parts. 5'. Long leg with sub-segmentation. 6. No flagellum. 6'. Flagellum.

carapace falls away and the limbs are slowly dragged from their casings. After a moult the animal is soft and temporarily defenceless, and is able to grow in size until the new exoskeleton hardens in its turn.

A valuable and characteristic feature of this method of growing is that it provides opportunities for the regeneration of lost limbs. Arachnids have the ability to cast a leg that has been injured or is trapped by a predator, and a new limb, appearing at the next ecdysis, then begins to form under the skin. The form which it then takes depends on the time that elapses between the accident and the moult. A limb may be replaced more than once, and more than one limb may be regenerated at the same time. The number of moultings between hatching and maturity depends on the size of the adult and may be as low as three or as high as fourteen. Some of the larger species, that live for several years, undergo a moult each year.

This power of limb-casting (autotomy) and subsequent regeneration is one of the arachnids' most valuable forms of self-defence. Other forms include the possession of strong, sharp jaws (chelicerae) which may or may not inject a poison with their bite, and the secretion of a malodorous fluid which repels the attacker. A few large spiders have the habit of scraping from the abdomen a shower of sharp spines or 'urticating hairs', which pierce the skin of the opponent and irritate like the sting of a nettle.

Throughout its life an arachnid, like every other animal, must feed itself, generally by catching and eating some animal smaller and weaker than itself. It either searches for these, usually at night, or lies in wait for such as are unlucky enough to approach. While the mouth parts cut and press the corpse, digestive enzymes are poured over it, and the resulting fluid is sucked into the gullet: thus digestion is actually external. An important feature of an arachnid's meals is that they do not have to be frequent or regular. The pre-digested fluid, when absorbed, can be stored in sacs leading from the gut and, in this way, relatively very large amounts can be preserved. The duration of the fast that they can survive is often remarkable. Some of the largest spiders have been known to live for two years without taking food or water. In general, a laboratory spider is overfed if it is given a fly every day.

When they reach maturity the males, even of the normally

sedentary families, wander in search of a mate, the finding of which seems to be at the mercy of chance alone. Their meeting is usually followed by a more or less prolonged series of remarkable actions, to which the name courtship has long been given and which varies from a stroking or tickling of the female by the male to a characteristic posturing and 'dancing' in front of her. There may be some form of actual union, or alternatively the male may produce a pillar-like spermatophore, carrying the sperm, and on to this he persuades the female to take a stand.

Like insects, arachnids are found in every country in the world: there are spiders and harvestmen in the Arctic, and mites in the Antarctic. There is scarcely an imaginable type of habitat which some at least have not colonized, and they spread from the highest mountains to the deepest mines, from low temperatures to desert heat and from sunlight to darkness. There are probably at least fifty thousand species, three-fifths of which are spiders, and new species are recorded every month.

Just as many people suffer from a deeply rooted fear of spiders, so there are a few zoologists who think that the Arachnida offer but a limited field for research. In refuting such a heresy I quote, by permission, the following lines from a letter written by a Sixth Form boy who had decided to produce a project on pseudo-scorpions. He wrote:—

"The trouble is I am getting carried away: I hate to think how long the project will be. As well as the straight stuff, there are all the insect associations, with ants, wasps, bees, flies, etc. etc.; with the symbiosis of ps.s on longicorn beetles, parasitism of bees, commensalism of ants' nests and predation of Collembola. They must be the only organisms to show the complete range."

I can wish the readers of this book nothing better than that they should acquire something of the same enthusiasm. Any who do so may be assured of a real reward.

3

The Order of Spiders
(Araneae)

The spiders are the most familiar, the most numerous and the most conspicuous members of the Arachnida in all parts of the world. They have always attracted the greatest share of attention, and this they have deserved since they are in every respect the dominant order of the class. They have received admiration, dislike, fear and even an undeserved neglect, and they have claimed places in myths, legends and superstitions to an extent that few creatures can equal. To the mind of a scientist their diagnostic features are:—

1 The cephalothorax and abdomen are united by a narrow waist or pedicel.
2 The carapace is uniform and bears six or eight eyes.
3 The abdomen carries a group of six or four spinnerets.
4 The chelicerae are pointed and contain poison glands.
5 The pedipalpi are leglike and carry the sex organs in the males.
6 Respiration is by lung-books or tracheae or both.

History Inevitably, spiders have found their way into the writings of naturalists from the time of Aristotle. We may mention the absence of any progress during the Dark Ages and neglect the earliest statements; indeed, the thirteenth century saw the rise of the myth of the tarantula and the remarkable effects of a bite from this actually harmless spider, belief in which lasted until the seventeenth century.

Spiders occupied the attention of an English physician, Martin Lister (1638–1712), who wrote the first account of British spiders in 1678. He described thirty-four spiders and three harvestmen. A more famous work was *Svenska Spindlar* by C. A. Clerck

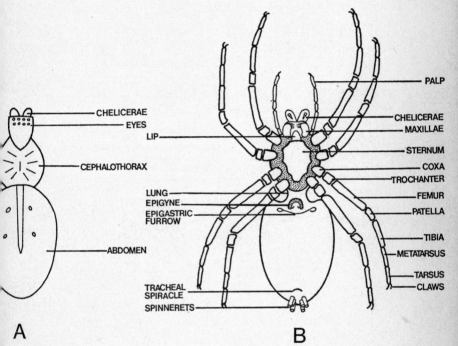

Fig. 2. Dorsal and ventral surfaces of a spider.

(1709–1765), published in 1757 and thus preceding by one year the tenth edition of Linnaeus's *Systema Naturae*. After this the study of spiders began to assume a truly scientific aspect.

The foundations of araneology were laid in France, where J. B. Lamarck (1744–1829), P. A. Latreille (1762–1833) and C. A. Walckenaer (1771–1852) all tackled the basic problem of classifying the increasing number of newly-discovered species. There was a contemporary interest in Germany, where C. W. Hahn

(1786–1836) and C. L. Koch (1778–1857) produced a sixteen volume work, *Die Arachniden*, with over five hundred coloured plates.

The greatest of all araneists, Eugene Simon (1848–1924) devoted himself to spiders from the age of sixteen and was the author of two standard works, *Les Arachnides de France* and *Histoire Naturelle des Araignées*, on which all later classificatory work has been based.

In Britain, John Blackwall (1790–1881) published *The Spiders of Great Britain and Ireland* in 1861–63, in which he described and illustrated over three hundred species. He was followed by the Rev. O. Pickard-Cambridge (1828–1917), rector of Bloxworth, who wrote *The Spiders of Dorset* in 1879–81; a book that retained its place as the only comprehensive work on our spiders until G. H. Locket and A. F. Millidge wrote *British Spiders* in 1951–53.

Pickard-Cambridge's position as our leading authority passed to Dr. A. R. Jackson (1877–1944), who, though unable to find the time to write a large book, was the author of many valuable papers and, above all, was the courteous and willing helper of all who sought his advice.

Scientific araneology was established in America, Africa and Australia later than in Europe, but is now studied there, and all over the world, with a growing intensity and realization of its value and attractions.

Classification The huge order may be divided into five sub-orders:—

Liphistiomorphae: primitive spiders, with a segmented abdomen, confined to Malaysia, Thailand, China and Japan.

Hypochilomorphae: a small group, in some ways intermediate between the above and,

Theraphosomorphae or *Mygalomorphae:* the trap-door and bird-eating spiders, of the hotter regions, but with two British species.

Gnaphosomorphae or *Arachnomorphae* or *Araneomorphae:* the largest sub-order, containing all the ordinary spiders.

Apneumonomorphae: another small group; characterized by the absence of lung-books.

The separation of about thirty thousand spiders into more than eighty families is clearly too large a task for a short introduction, where a simpler treatment is appropriate. If we glance back at Blackwall's book we find his species arranged in ten families whereas Locket and Millidge put nearly six hundred species into twenty-three families. The increase is not wholly due to new discoveries, but is rather the result of a more meticulous examination of spiders' bodies and the use of fresh characteristics for the purpose of classification enabling a large family to be divided into several smaller ones. Details of these changes may be read in Locket and Millidge's books. For the present, Blackwall's ten families will be used to introduce beginners to the chief types of spiders they will find here.

1 *Atypidae*. This is a family of the Theraphosomorphae, with two species of the genus Atypus found in this country.

2 *Lycosidae*. These are the wolf-spiders, the fairly large brown spiders sometimes to be seen in large numbers running in woods and fields. About forty species are listed and cover nearly all possible habitats from mountains to the sea. They are keen-sighted, active spiders: the females carry their cocoons attached to their spinnerets, and the newly hatched nymphs ride on their mothers' backs.

3 *Salticidae*. These are the jumping spiders, with very large median eyes, keen sight and the habit of leaping upon their prey from a distance. Many of them are brightly coloured, and are attractive creatures when they are seen hunting in the sunshine in a glade in the woods. Their elaborate courtship dancings were the first examples of such behaviour to be described among spiders.

4 *Thomisidae*. These are the crab-spiders, with strong anterior legs, stretched out sideways and so enabling the spider to run in a crab-like fashion. Some of them lurk among the petals of flowers, with the colours of which they blend. Others, the Philodrominae, stretch their legs fore and aft as they lie along a blade of grass.

5 *Gnaphosidae*. This family and the related Clubionidae contain spiders that generally hunt on the ground by night, though some of them climb at certain seasons into the shrubs and trees. They are usually soberly coloured spiders with smooth, shining

abdomens resulting in the name 'mouse-spiders'.

6   *Ciniflonidae*. Spiders of this family possess the modification of a pair of spinnerets into a perforated plate, the cribellum ('small sieve'). The silk, which is combed from it by a row of setae on the fourth legs, gives their webs an unmistakeable bluish appearance. Some of our largest species belongs to this family.

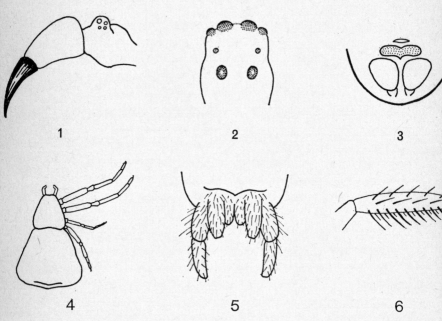

Fig. 3. Features of six spider families.

1. Chelicera of Atypus.   2. Eyes of Salticidae.   3. Cribellum.   4. Legs of Thomisidae.
5. Spinnerets of Agelenidae. 6. Tarsus of Theridiidae.

7   *Agelenidae*. The familiar cobwebs of our houses, spun by the members of this family which can be recognized by the length of two of the spinnerets.

8   *Theridiidae*. A large family of often gaily coloured and generally small spiders, which live among the leaves of shrubs and bushes, where they spin webs of an irregular pattern. The

notorious Black Widow belongs to this family. All species have a characteristic comb on the tarsi of their fourth legs, with which they draw out a ribbon of silk to be thrown over their captives. Some species feed their young at their mouths.

9 *Linyphiidae.* The small black 'money-spiders', which are held to foretell a promise of riches, constitute the bulk of this family. It is by far the largest of the British families, with over two hundred species, many of which are small, black and without a pattern. Others are to be seen hanging upside down under a sheet-web in almost every hedge. The family dominates the spider population of Britain, especially in the north, whence it extends into the arctic regions. The small size of the species makes a microscope necessary for their identification.

10 *Argiopidae.* The circular orb-web, so often to be admired, is the work of the garden spider, *Araneus diadematus*, the commonest member of a family characterized by this form of web. This is the most highly specialized of all spider families and some of its exotic species spin huge webs of surprising strength. There are nearly forty British species, most of them bearing a colourful pattern on their backs.

Of the nearly six hundred British spiders, about one hundred and twenty may reasonably be described as 'common', 'widespread' or 'abundant' whilst the rest are either locally distributed, or are only occasionally seen because they are well concealed. A few are such rarities that they have been reported only a few times in the present century. Much encouragement follows the information that intelligently-directed searching may expect rewards in the form of some of the more elusive species.

A characteristic feature of the British fauna is the high proportion of the members of the families Linyphiidae and Erigonidae to be found. Serious araneists cannot evade the task of identification which these small spiders present for they are superficially so much alike that careful scrutiny is necessary to determine their names. Some naturalists genuinely enjoy a challenge of this kind and derive a well-deserved satisfaction as each problem yields to their efforts; others, of which I am one, do not do so. Having found myself as a beginner in a district well-populated with members of the genus Lepthyphantes, I developed a prejudice against the whole family, which has lasted for more than fifty years. Wiser zoologists should not allow themselves to acquire this limitation.

The Living Spider Like every other arachnid, save the scorpion, the spider hatches from an egg, laid with a number of others in a silk cocoon and usually casts its skin for the first time while still within the egg membrane, which it breaks with the help of an egg-tooth on its infant chelicerae. It leaves the cocoon as a nymph and starts an independent life.

The cocoons of spiders are similar in general plan and are made by depositing eggs on a silk mat and covering them with a silk sheet with, occasionally, a layer of floss between the egg mass and the outer coat. Treatment of the cocoon is variable. Most of them are fixed to the underside of a sheltering stone or are left under the bark of a tree or attached to a wall or fence. Web spiders usually hang their cocoons near their webs. A few show some degree of maternal care, and carry their cocoons with them, either attached to their spinnerets, as in the Lycosidae, or held uncomfortably under the sternum, as in Pisaura.

Some young spiders, and a few older ones, adopt the method of dispersal known as gossamer, in which they float through the air on a long strand of silk. Their setting forth on these chance journeys may be seen in spring, when on warm days, when a current of air is rising from the ground, the spider climbs to the top of the nearest eminence, turns up-wind, raises its abdomen and secretes a drop of silk. The breeze draws this out until the buoyancy reaches a point at which the spider lets go and sails away. These migrants have been known to reach heights of ten thousand feet and to travel over two hundred miles.

The growth of the young nymph is a discontinuous process, already mentioned in Chapter 2, where its association with the regeneration of lost limbs is seen to be an important contemporary. Among spiders the task of ecdysis is a long and tedious one, especially in its later stages. The withdrawing of the eight legs simultaneously from their old sheaths is achieved by a series of steady heaves, more than a hundred of which may be needed.

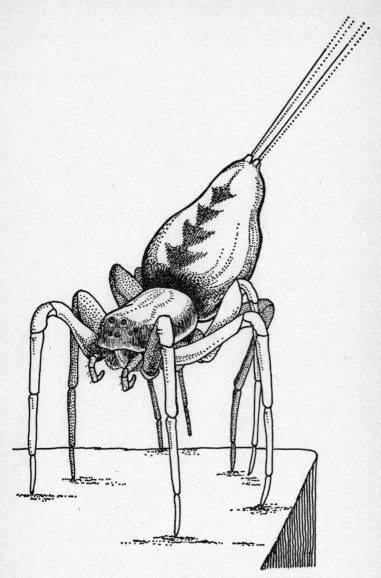

Fig. 4. Spider preparing for gossamer flight.

Careful preening follows liberation, and has the result of setting all the sensory setae in the correct directions.

The shortened survey of the families of spiders earlier in this chapter should have made it clear that many families have never adopted any more elaborate way of feeding themselves than a general search during their nightly wanderings. Others hunt with a greater activity, either pursuing their prey or leaping upon it: others again lie in wait, often more or less concealed by the blending of their colours with the leaves or petals among which they live. Such protective colouring lessens the inevitable risks from predators, but protective resemblances, such as are seen in stick-like larvae and leaf-like insects, are rarer. The most surprising exception is the eastern spider, *Phrynarachne decipiens*, which, resting on a mat of silk looks like a bird's dropping.

Surpassing all these features of a spider's life is their wholly characteristic and wonderfully successful use of silk. With no more than an excusable exaggeration this has been described in the following words:—

"The young spider is born into a silk nursery, on a silk monoplane it flies away; with a silk web it catches its food, binding up with silk threads and ribbons its struggling prey or its bitter enemies. It drops from peril on a silk rope, of a silk sheet it makes its cocoon, its eggs wrapped round with silk cushions. In a silk chamber it sleeps through the cold of winter, and even in death it is sometimes wrapped in a silk shroud."

It is, in fact, literally true that many a web-spinning spider is never out of contact with silk throughout its life, a phenomenon to which Tilquin has given the name of *sericiphilie*.

The web is, however, the main product of the spider's exploitation of silk. Usually the web is peculiar in form to each family or group of families; for example, everyone can recognize the orbweb spun by the Argiopidae. The cobwebs of our houses and sheds are derived from a tubular resting place, one part of which is extended into a sheet of greater or lesser dimensions, and the passing insect is likely to collide with threads above this and to fall into its entangling meshes. Other webs are horizontal sheets or hammocks, under which the spider hangs upside down, and others again seem to be no more than irregular tangles of threads, stretched in all directions. Actually, there is more evolutionary relationship between the different forms of web than is apparent on the surface, and they provide an admirable example of adaptive radiation.

The spider's sex life has long been a topic of surprise and popular mistakes. Mature male spiders that have been accustomed to leading the usual sedentary life in a web, cease with their maturity to spin and take to a wandering life, during which they may come upon a female, either on the ground or in her web. Their approach is always a signal for the commencement of the curious series of instinctive actions which are described as courtship. This may consist of a mere mutual stroking of the legs, or of a display of decorated legs and palpi, or of a more elaborate dance by the male in view of the female. Among web spinners, the male either plucks at the threads of the web or drums upon it with his palpi. These preliminaries may lead to the fertilization of the female.

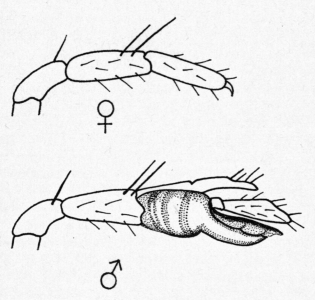

Fig. 5. Male and female pedipalpi.

Since the sperm are secreted from glands in the abdomen, a male has first to charge the organs on his palpi with the seminal fluid. This is deposited on a special sperm-web and absorbed from it by the palpi, which are then applied to the female's epigyne.

There is a widely held belief that the female 'always' slays and consumes the male. This is untrue; usually the male moves away as soon as the mating is accomplished. In the abnormal circumstances of mating under the eye of an arachnologist and in the confines of a small cage, a hungry female is likely to indulge in this peculiar form of cannibalism, but to accuse the whole order of such behaviour is only a popular exaggeration.

Opportunities There is still much to be discovered about the day-to-day habits of many of the British species. The capture of prey and the treatment of the victims is reasonably well known for the orb-spinning Argiopidae, for Atypus, and for some of the Theridiidae and Linyphiidae. Something, too, has been written about the hunting of the Lycosidae and the leaping of the Salticidae. But among the less spectacular families our comparative ignorance is too great for complacency; it argues an indifference that is surprising and even reprehensible. Spiders belonging to the families Dictynidae, Gnaphosidae, Clubionidae, Thomisidae and others fall into this category. In fact, it is probably no great overstatement to say that every such species has some detail to reveal to careful observers.

Early in the morning or later, in the evening, the garden spiders spin their webs, and anyone who chances to see one so occupied should stay and watch the process to the end. It is an example of instinctive actions that has been carefully watched and photographed and analysed and discussed at great length, especially during recent years. But this concentration on the orb-web has left web-watchers (if there are such persons) with the opportunity to watch webs of other designs, and to try to detect the principles that guide their construction. In 1970 Dr Lamoral made the striking discovery that the apparently fortuitous tangle of threads in the Theridion web has grown systematically from four original straight lines. He coloured these four distinctively as they were spun, and was able to watch their varying fates as the rest of the web grew round them. Enthusiasts should be encouraged to watch other webs, with the same intention in view.

When it is realized how easily spiders can be kept for observation in cages, a wide field of possibilities is opened. Many zoologists who have kept spiders in large numbers for considerable periods of time have described how the animals would live placidly on a slip of wood in a test tube closed with a cotton wool plug, and that they would periodically accept a fly presented to them in forceps. But what a caricature of their natural lives is such a set-up. It seems to be axiomatic that a spider under critical observation should be given conditions which as nearly as possible resemble those in which it has been accustomed to live.

This gives scope for originality in the construction of vivaria in which the spiders are kept. Rocks may be represented by large stones, sand can be used where sand is a natural part of the environment, vertical surfaces can be roughened with sheets of glass paper, water can be constantly available. Any cage should be

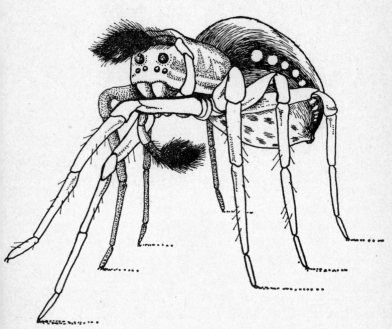

Fig. 6. The courtship of the male wolf spider.

as large as possible. A house spider will, indeed, survive in a jam pot, but a normal web will not be spun in a circle, and a *Tegenaria* should be given at least the latitude of a shoe box.

The courtship of spiders may reasonably be watched, and attention may be directed towards a decision of the meaning or 'purpose' of courtship in spiders and in other arachnids. Zoologists who have studied the phenomenon in other animals have detected five results of animal courtship in general, namely that it:—

i   Brings the sexes together at the right time and right season
ii Ensures that mating occurs where the young will find their needs met
iii Excites both sexes and represses aggressive tendencies
iv Prevents hybridization
v   Reduces competition from other males.

Without denying the truth of any of these conclusions, an arachnologist should try to assess their relative importance among Arachnida.

Finally, I would mention two small mysteries that have been with me for some years.

In 1931 I was intrigued by the fact that two closely related species *Zygiella atrica* and *Zygiella x-notata* were to be found respectively on bushes and on walls or windows, and neither in the habitat of the other. Experiments that I made suggested to me that this was due to a preference by one species for a dry and by the other for a humid atmosphere, and I coined the word 'hygrotropism' to describe this form of behaviour. My conclusions have since been denied. The construction of efficient choice-chambers has improved immensely since that time, and the work might very well be repeated.

A second problem refers to the common species *Agelena labyrinthica*. I have found it to be invariably in its webs on gorse bushes and in no other kind of plant, and when I have said so, the invariability of the choice has been denied. Yet I remember walking one day up a road in Northern Ireland where gorse and other shrubs were intermingled on both sides, and seeing *Agelena*

Fig. 7. The web of the *Zygiella*.

on nearly every gorse bush and never on any of the others. Here, too, more investigation would be interesting.

It seems to be undeniable that any experimentally-minded araneist has plenty of problems at his disposal.

4

The Order
of Harvestmen

(Opiliones)

The long-legged harvestmen may justifiably claim to deserve the notice of arachnologists more strongly than even the spiders. They are more easily named, there are fewer species in Britain, they live comfortably in the laboratory, and, since they have attracted but little attention, they offer greater opportunities for original observations. Their diagnostic features are:—

1   The body, often oval in outline, has cephalothorax and abdomen joined across their whole breadth; there is no pedicel.
2   There are two simple black eyes, mounted back to back on a raised ocular tubercle.
3   The legs of most species are disproportionately long.
4   The pedipalpi are leglike, and do not act as pincers.
5   A pair of odoriferous glands are present in the cephalo-thorax.
6   The abdomen of the male contains an extrusible intro-mittent organ.

History   Although the name Phalangium was used by the ancient classical writers, the word did not imply a distinction from spiders.

In 1758 Linnaeus made it the name of one of his genera of apterous insects; it contained three species, one of which was the common European *Phalangium opilio*.

In Britain the first description of native species was a paper by R. H. Meade (1814–1899) printed in 1855. It named fifteen species, and was followed by a supplement in 1861. The seventh volume of

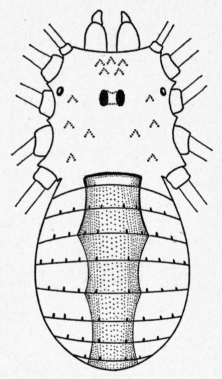

Fig. 8. A harvestman.

Simon's *Arachnides de France* contained two hundred pages on the order, with its first satisfactory classification. The British species were monographed by O. Pickard-Cambridge (1828–1917) in 1890, and an immense thousand-page volume, *Der Weberknechte der Erde*, was published by C. F. Roewer (1881–1965) in 1923. Five years later the first attempt to describe the general habits of harvestmen was written by an Austrian zoologist, H. Stipperger.

The distribution of harvestmen in Britain was summarized by Bristowe in 1949; valuable ecological studies were published by V. Todd in 1949 and 1950, and their physiology was investigated by J. Phillipson in 1959 and 1960. Today the leading opilionologist is Prof. C. J. Goodnight of Kalamazoo, Michigan.

Classification. The order is unequally divided into three sub-Orders:—

    1 *Cyphophthalmi*, the most primitive sub-order, occurring in well-separated localities, chiefly in the tropics. A few species are Mediterranean, some occur in the south of France and others in New Zealand.

    2 *Laniatores*, including the most specialized families. Nearly a thousand species are known, chiefly from the southern hemisphere. One family, the Phalangodidae, is European.

    3 *Palpatores*, about equally numerous, are the dominant group in northern temperate lands and are not as widely distributed in hotter countries.

There are twenty-four British species, all of which belong to the third sub-Order which is divided into two super-families; the *Eupnoi*, including the Phalanglidae family, in which the tarsus of the palp is longer than the tibia and carries a claw, and the *Dyspnoi*, in which it is shorter and has no claw. The latter includes both the Trogulidae and Nemastomatidae families.

The Living Harvestman. The eggs of harvestmen are usually laid underground in clumps of about twenty; they are pale yellow

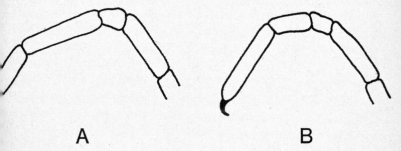

A   B

Fig. 9. Pedipalpi of A Dyspnoi and B Eupnoi.

spheres, less than half a millimeter in diameter, and in this country they hatch in about three weeks. The very young harvestman is a feeble little thing, which seems to have some difficulty in managing its long legs. It is not much more than an hour old before it casts its skin and moults for the first time. The nymph is a fascinating little creature; for full appreciation of its detail it should be looked at through a microscope.

If food is available, moulting, or ecdysis, will occur at intervals of about ten days, but often the attempt to raise a young harvestman to maturity fails after about its third or fourth moult. This seems to be due to the continuing moistness of the atmosphere in the cage, which prevents the new exoskeleton from hardening properly. To avoid this, the lid should be raised from the cage for a period every day, a 'change of air' being apparently as beneficial to young harvestmen as to young children. From seven to ten moultings bring the harvestman to maturity.

Its diet during this time is varied. Normally it hunts at night and picks up the insects and other small animals that it meets, seizing them in its jaws and evidently possessing enough strength to kill them without the help of poison. Those that are recalcitrant are sometimes treated to an unusual type of behaviour: they are surrounded by the eight legs as in a palisade while their captor falls upon them from above like a minute pile-driver. As is mentioned elsewhere, harvestmen also eat dead matter, both animal and vegetable, and drink fruit juices from fallen bruised fruit as well as water.

Harvestmen suffer less from the cold than they do from thirst. If an over-wintering specimen is extracted from beneath the snow it seems to be quite lifeless, but a few minutes' warmth restore it to summer-like activity. Thirst, if prolonged for more than a few days, makes it stiff and torpid, but again an offer of water is greedily accepted and normal health soon restored.

During its life a harvestman is the possible prey of a few natural enemies, such as are accustomed to feed on insects. Some of these, however, including a number of spiders, reject a harvestman almost as soon as they have picked it up. This is due to the harvestman's effective method of defending itself by the device of the gas-bomb. In the forepart of the body, just above the origins of the second pair of legs, there is a pair of odoriferous glands. In circumstances of stress or threat these glands emit a vapour (or sometimes a jet or spray) of peculiarly-smelling liquid, from which the predator turns away, presumably in disgust. Its scent is not always very apparent to ourselves, and varied descriptions have been given of its nature.

At frequent intervals, and especially after moulting, a harvestman preens itself in much the same way as do the spiders described in Chapter 3. A difference is due to the disproportionate length of a harvestman's legs for, as a leg is drawn through the chelicerae, it bends almost into a circle until the claw is reached, when it escapes and shoots out like a released spring.

The legs of harvestmen are shed even more easily than are the legs of spiders, whenever the sacrifice allows the owner to escape from danger on the legs that remain, and here again harvestmen are unlike most of their relations. Sometimes when the harvestman is drinking, a leg immersed in the water may be held by the surface tension so securely that the animal cannot pull it out, but casts it off instead. On rare occasions this capture of a leg results in the harvestman's being drowned—one of the few ways in which natural death is not the consequence of capture by a carnivore. Another difference from other arachnids is the fact that a leg so autotomized is not reproduced on moulting. Combined with the ready loss of a leg this is particularly surprising, and yet towards the end of the year a harvestman still in possession of all its eight legs is comparatively uncommon. There is a well-known record of a harvestman seen running in the field with only two legs.

Like so much else in their lives, the sex life of harvestmen is very characteristic. There is no preliminary courtship, such as gives the behaviour of spiders and scorpions so great an interest: the sexes unite as soon as they meet. The male has the peculiarity of an intromittent organ ensheathed in the abdomen, and this is passed out, through the chelicerae of the female and inserted below the genital operculum. Mating lasts but a few seconds, and may be repeated often by both partners, not necessarily with each other. The female has a long, elastic ovipositor, and in due

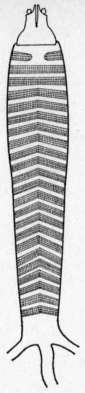

Fig. 10. Ovipositor of Harvestman.

time she thrusts this under the soil to deposit the eggs, in which she takes no further interest.

Opportunities Because they are very easy to keep in cages the habits and behaviour of harvestmen are wide open to observation, and there is no doubt that surprises are awaiting discovery. Moreover, the nocturnal disposition of harvestmen means that they may be most profitably watched after supper, when the day's work is over. Observation may well continue into the night, for such vital activities as moulting and egg-laying are usually carried out later than sunset. The way in which the harvestman hangs itself up to moult, or the way in which it moults when, as sometimes happens, it moults 'standing up', always deserve full description.

It may be that the animals appear to be affected by the light necessary to the observer. One way to meet this difficulty is to replace the ordinary light bulb by a red one, for harvestmen, like many other invertebrates, are relatively insensitive to red rays. Even if the observing is less satisfactory, it is better than having nothing to observe.

It is a fact that most of our knowledge of the daily habits of harvestmen has been based on the common species, *Phalangium opilio*, *Leiobunum rotundum*, or *Oligolophus agrestis*. An ambitious observer would find great opportunities in similar keeping of and observing the day-to-day behaviour of some of the other species, such as *Mitopus morio* and *Oligolophus spinosus*. There are certain to be differences, and 'comparative opilionology' presents a wide field. Even more interesting would be a similar series of records of species found in southern France or in the Mediterranean countries. Holidays may be profitably exploited to obtain specimens.

Mention has been made of the unusual reproductive activities of harvestmen, and this recalls the long-established remark that the males indulge in 'combats acharnés' or 'bloodless battles' if two of them meet in the presence of a female. This unexpected detail has been briefly denied, but not critically examined. It would be well worth while to keep four or five males in a cage for a few days, then one evening to introduce a female, and watch the consequences. The common species *Phalangium opilio* would be suitable because the horn on the chelicera makes the sexes distinguishable at sight. The real nature of the competition, if any, ought to be better known than it is at present.

A curious feature of harvestmen, that has been noted by several writers and fully explained by none, is their effect on one another when several are kept close together. I noticed it first when it was my custom to keep an empty jam pot at hand while I was working in my garden: into this I dropped all the harvestmen that I disturbed. By the end of the morning the catch of a dozen or so were completely inert, as if under the influence of an anaesthetic. From this they recovered soon after they were separated. Some explanation of this is needed.

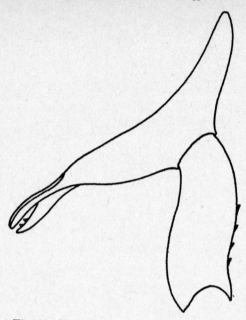

Fig. 11. Chelicera of male *Phalangium opilio*.

One of the most intriguing mysteries of opilionology at the moment is the occurrence of occasional parthenogenesis. A very attractively coloured species is *Megabunus diadema*, a beautiful little green, white and silver species, more often found on hills than at lower levels. Collectors take the females in considerable numbers, but have always found the males to be almost vanishingly scarce. The proportion of one to 407 has been quoted. We are told that the survival of the species is due to the unique ability of the female to lay eggs that develop normally without the customary stimulus of fertilization.

A fact that makes this belief more interesting is the support it receives from some observations of eggs laid by females of other species, in which the males are normally frequent. Three females, known to be maidens, have laid eggs, a small proportion of which have shown partial development. The early divisions of the ovum have been followed by an approach to the later stages, and although no hatchings were recorded the tendency has been undeniable. There is room for advance here.

5

The Order
of False-scorpions

(Pseudoscorpiones)

The little false-scorpions form one of the most fascinating orders of small animals that are to be found anywhere; indeed it is difficult to understand why they are not among the most popular objects for naturalists to collect and study, since some zoologists, eminent and amateur, have been attracted to them by their first sight of a living specimen. They look very much like scorpions without tails, and their diagnostic features are:—

1 The cephalothorax is covered by an undivided carapace, with 4, 2 or 0 eyes.
2 The abdomen consists of twelve distinguishable segments.
3 The chelicerae are small pincers, containing glands that secrete silk.
4 The pedipalpi are enlarged, like those of real scorpions, and contain venom glands.
5 Legs i and ii are usually different from legs iii and iv.
6 There are two pairs of respiratory spiracles, on abdominal segments 3 and 4.

History Aristotle wrote of 'the creature like a scorpion found in the pages of books', and 'it is similar to a scorpion but without

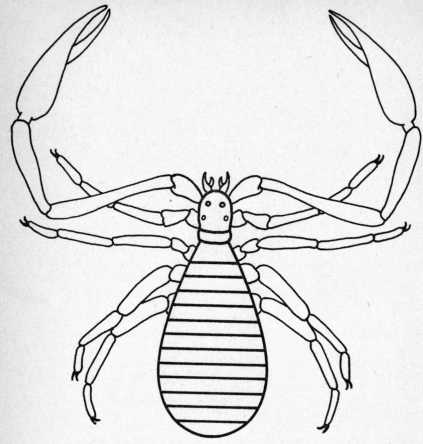

Fig. 12. A false scorpion.

a tail and is very small'. Linnaeus in 1758 listed *Acarus cancroides* and *Acarus scorpioides* in one of his genera. The first zoologist to give false-scorpions fair treatment in English was W. E. Leach (1790–1856), who described eight species in two genera, Chelifer and Obisium, based on a collection of 18 specimens which is still preserved. His work was followed by H. W. Kew (1868–1930), who was the first widely recognized authority on the group, to which he made notable contributions.

In 1920 J. C. Chamberlin (1898–1962) began his life-long study of the order, receiving specimens from all over the world and constructing a comprehensive classification, published in 1931. Contemporary with Chamberlin was Max Beier of Vienna, who, though professionally an entomologist, was also an authority on false-scorpions. He wrote over one hundred papers on them, and one of the most pleasing features of his work was his amiable relations with Chamberlin. The two zoologists freely exchanged their discoveries and opinions in a spirit which is one of the outstanding characteristics of all arachnology. At present the biology of false-scorpions forms a prominent part of the work of P.D.Gabbutt (Manchester), M. Vachon (Paris) and P. Weygoldt (Freiburg in Breisgau).

Distribution About two thousand species of false-scorpions are known. They are to be found everywhere, being completely cosmopolitan with the exception of the frigid zones of the Arctic and Antarctic. They lead the strictly cryptozoic life characteristic of primitive orders, but sometimes appear in our houses.

Classification Three sub-Orders are recognized: Heterosphyronida, Diplosphyronida and Monosphyronida, with twenty families in all. The first of these contains the family Chthoniidae, with six British species of the genus Chthonius; the second includes the family Neobisiidae, six species, and the last the Cheiridiidae, one species, the Cheliferidae, three species and the Chernetidae, ten species. All particulars necessary to the British arachnologist are to be found in the Linnean Society's Synopsis of the British Fauna, No. 10, by G. Owen Evans and E. Browning.

The Living False-scorpion First acquaintance with a false-scorpion most often comes when one of them, sifted from dry leaves, falls on to the receiving sheet. While many animals run rapidly away, a false-scorpion stays still and, with its legs drawn in, appears as a squarish object, which soon becomes unmistakeable. After a few moments it moves, walking slowly with a characteristically dignified air of extreme deliberation, unlike that of any other creature. One is immediately struck by this demonstration of character in a small animal which displays in its bearing the equivalent of human personality. Small wonder that zoologists are apt to be immediately attracted.

As it thus proceeds, the large pedipalpi are stretched out in front, their long setae acting as sense organs. Should it make a disturbing encounter it withdraws its palpi and runs rapidly backwards, and in this characteristic response does not look like a false-scorpion at all.

In this manner false-scorpions hunt at night. Unlike harvest-men they seem never to eat dead matter, but seize their prey with their pedipalpi, killing their victims with venom, as do spiders. They exhibit a ferocity that is some compensation for their small size and I have seen a spider die instantaneously when bitten by a false-scorpion smaller than itself.

As a preliminary to a meal a false-scorpion often rubs chelicerae and pedipalpi together, whereby the latter cleans the grooves and channels in the former, so that absorption may be uninterrupted. A false-scorpion may often have dirty palpi, but its chelicerae are always clean.

The digestion of the food begins, as is common among arachnids, outside the body. The food particle is held by the chelicerae, glands pour enzymes upon it, and the fluid thus produced is sucked into the stomach. A surprising feature of the feeding is the complete consumption of the food since, among spiders and harvestmen, there is a dry residue of 'left-overs', but a false-scorpion leaves hardly a trace. Meals may be given to captive false-scorpions at intervals of a few days. H. W. Levi has reported that his false-scorpions fed once a week were in better condition than those captured in natural circumstances.

The chelicerae have a dual function, for they secrete silk, which is used in the construction of protective cocoons. In these moulting takes place, eggs are laid and the bringing up of the family is begun.

The construction of these 'cocoons' or rest-chambers is carried out with an instinctive precision which equals that of the spider's spinning of an orb-web. After seeming to choose a suitable spot, the female collects small pieces of grit and wood and arranges them in a circle round herself. Upon these she places a second row, securing them in position with threads of silk from the cheliceral glands. As she continues in this way the wall rises, leaning inwards towards the centre and being constantly reinforced by the brushing of silk on its inner surface. The collecting of more stones involves the maker in stepping over the wall, and this she does until the hole left at the top is too small. The placing of a final piece of grit completes the igloo-like structure and imprisons the builder. She continues to brush the inside with silk until, after a few days, its strength is assured.

Within its shelter the false-scorpion moults, a process that is a quieter one than the struggling of a spider, and which occupies the best part of a day. In the course of its life a false-scorpion moults four times before it is fully grown. It may live for three or four years, passing the winters in a similar shelter, which may be shared with two or three individuals.

The courtship of false-scorpions is a remarkable performance. The male, on approaching a female, shakes his abdomen and waves his pedipalpi. If the female does not move away he produces a structure known as a spermatophore. This is dropped from his genital orifice as a vertical thread, which quickly hardens.

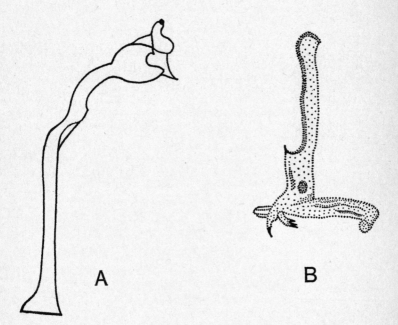

Fig. 13. Spermatophore of A scorpion and B false scorpion.

Upon its apex he deposits a packet of spermatozoa and retreats. Thereupon the female advances and stands over the spermatophore, which thus enters her body. The male returns, seizes her legs and shakes her, ensuring the detachment of the sperm packet and the liberation of the spermatozoa within her.

A really remarkable feature of false-scorpion reproduction is the occurrence of fertilization without courtship or mating. The males of some species may produce a spermatophore although no female is present, and abandon it. It may be found by a wandering female, who then straddles it and departs with its valuable packet. Thus is all romance taken from the inauguration of a new generation.

In due course the fertilized female builds herself another cocoon and in this she produces a brood chamber. This is secreted by the walls of the oviduct in the form of a mushroom-shaped container, into which a score or so of eggs are now passed. The brood chamber remains attached to the female's body. When the young break out of the egg-membranes they remain in communication with their mother. Yolk, 'false-scorpions' milk', is forced into them, and as they grow the brood chamber increases in size and must be a serious inconvenience to the mother. Finally the young ones moult, sever their attachments from their mother, and leave the brood chamber to start an independent life.

Opportunities It is clear that this surprisingly complex life-history offers countless opportunities to every arachnologist who is attracted by watching the details of the lives of these animals, and their small size presents difficulties which make the challenge more compelling.

One of the characteristics of false-scorpions is the existence in the males of remarkable objects known, from their shape, as ram's horn organs. These are evident at the time of courtship, when they are probably the source of some odorous secretion. Observers might well look for these curious secondary sexual characters, and hope to notice some active part played by them.

False-scorpions, more than any other common arachnid, offer to the microscopist innumerable opportunities for the exercise of his skill. Chamberlin's method of mounting a specimen in such a way that the characters of the species are displayed is described in Chapter 9. The arachnologist who wishes to learn more about the structure of a false-scorpion's body would do well to concentrate his attention on the chelicerae. These must rank as among the most versatile organs in the arachnid world.

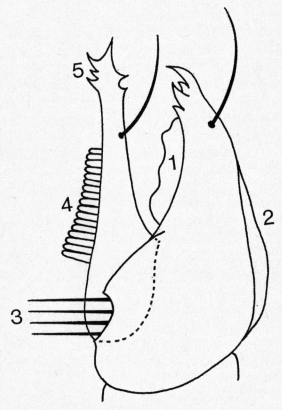

Fig. 14. Chelicera of false scorpion.

1. Lamina inferior.   2. Lamina exterior.   3. Flagellum.   4. Serrula exterior.   5. Galea.

Primarily, they accept captures from the pedipalpi and are responsible for tearing open the body and injecting the digestive enzymes. They also secrete silk, used in the making of the rest shelters, and are provided with sensory setae on both fixed and moveable fingers, the number of which serve as some indication of the age of the specimen. Further, the chelicerae carry a group of very specialized blade-like setae, known as the flagellum, varying in number from twelve to one. They are used to clean the mouth parts after a meal, and their use should be looked for. Details such as these, and the form of the spinneret, are minutiae that must delight the heart of any inquisitive microscopist.

At certain seasons of the year false-scorpions of a few species may be found holding with their chelicerae to the legs of insects or harvestmen. This is the phenomenon of phoresy, the chief mystery of false-scorpion biology. They are not parasites, and all that they seem to gain is free transport. Arachnologists should seek, especially in August, for false-scorpions in these positions: a harvestman has been known to be carrying eight passengers.

Almost without exception these are mature females, and since they are found in this situation during the summer, when flies are abundant and the false-scorpions are pregnant, the first suggestion is that the habit is of value because it disperses the species. Other suggestions have found support. Many phoretic specimens have been shown to be in an underfed condition, and are using their carriers to take them to richer feeding grounds. Another view is simply that the false-scorpion has seized the leg of an adjacent insect or harvestman in its normal search for food, and had been transported by accident. The subject is one that deserves any attention that observers are willing to give it, particularly with respect to the habits of the transported species and their likelihood to meet possible carriers.

6

The Order of Scorpions
(Scorpiones)

Although scorpions do not live in Britain, no book on arachnology could deny them a section, for they have too much importance for all zoologists. Moreover, the millions of our countrymen who now annually take their holidays in Spain or Italy are likely to meet scorpions and to find an interest in them. No one could possibly mistake a scorpion for any other creature, and their diagnostic features are:—

1   The abdomen is clearly divided into a broad mesosoma of seven and a narrow metasoma or 'tail' of five segments.
2   The last segment carries a venomous sting.
3   The carapace, which is unsegmented, carries two median and two to five pairs of lateral eyes.
4   The pedipalpi are enlarged, powerful chelate organs.
5   The second abdominal segment carries a pair of combs or pectines.
6   There are four pairs of lung-books.

History   The scorpion has always been so familiar that four thousand years ago the Chaldean astronomers gave it a place in the Zodiac. It was mentioned in very early Chinese writings and in Egyptian mythology was associated with the goddess Isis and in

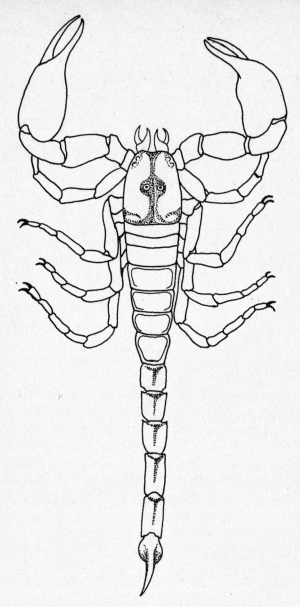

Fig. 15. A scorpion.

Persia with the god Mithras. Aristotle and Pliny wrote fully about scorpions.

The first scientific account of scorpions was written by Maupertius (1689–1759), a French mathematician; and in 1758 Linnaeus named five species of scorpions in the *Systema Naturae*. Later the scattered knowledge of these animals was brought together by E. Simon (1848–1924) in 1879; their classification was established by Kraepelin in 1899. Henri Fabre wrote several essays on the species *Buthus occitanus*.

Between the wars great progress was made in the study of scorpions both in Europe and America. The nature and effects of scorpion venom were investigated and antidotes produced. Most recently knowledge of the African scorpions has been advanced by the labours of R. F. Lawrence and M. Vachon.

Distribution Scorpions are confined to the warmer parts of the earth, and can live in the heat of the deserts. In Europe they occur in the south of France and Spain, as well as in Italy and Greece and all the shores of the Mediterranean. They cover Africa, and in America reach as far north as Oregon. They are widespread south of the equator, but are absent from New Zealand and the oceanic islands.

Classification Scorpions have a fairly full fossil record, and experts have divided them into nine groups or 'super-families', from only one of which, the Scorpionoidea, all living scorpions have descended. This points to much unsuccessful experiment in the earliest days of scorpion history. The survivors are divisible into seven families, containing in all about ninety genera and over six hundred species. Of these more than three hundred are placed in the family Buthidae and about a hundred and fifty in the Scorpionidae.

THE LIVING SCORPION

THE LIVING SCORPION   Scorpions in general pass the day in hiding, safe from their enemies and, more importantly, from the drying effects of the heat of the sun. Their hiding places are determined by the nature of their environments, for there are scorpions of the desert, scorpions of the forest and scorpions of moist, if not marshy, areas. The desert scorpions have the most difficult time. Many of them dig efficiently: they scrape away the soil with their palpi and legs, sometimes making a shallow trench by the side of a stone, sometimes digging deep burrows some of which may go down two or three feet, and the scorpion may remain in it for days at a time. They may come periodically to wait at the entrance, on the alert for passing insects. Scorpions in forests and jungles can usually find shelter under the loose bark of trees or logs, whilst others creep under stones which provide more effective protection than might be expected.

Sometimes, when scorpions are unavoidably exposed to heat they adopt a habit which has been described as stilting. They straighten their legs and stand raised, so that their bodies do not touch the ground and a current of cooling air can circulate below them.

When abroad at night, scorpions seldom search for their prey, but wait, almost like spiders in their webs, for wanderers to approach. Evidently they are made aware of this by the fact that the movements of the victim set up vibrations in the air which are received by the long setae on the pedipalpi. Usually the scorpion then seizes its prey with its great chelicerae, holds it aloft and brings the sting forward over its back to paralyse it. The chelicerae dismember it and the stomach sucks it in. A meal may occupy an hour or more.

Starvation is of little importance to scorpions, and desert scorpions get the water they need from their food. Others drink freely: they can collect drops of water in their pedipalpi and carry them to the chelicerae. In captivity, scorpions drink daily.

The danger of the scorpion's sting is generally exaggerated, for the sting of most species is not much more painful and is no more dangerous than the sting of a wasp. However, a matter of some dozen species have been shown to be virulent. Like some other venomous animals, scorpions have the curious habit of stridulation, in which a peculiar purring noise is produced by the rubbing of a scraper, consisting of short, stout tubercles, against a rasp, a roughened surface. Usually the scraper is on the first segment of the pedipalp and the rasp on the chelicera or leg; but in some species these positions may be reversed. It is usually supposed

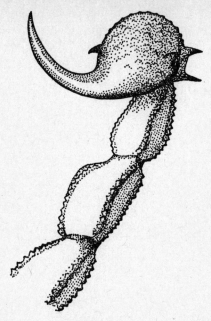

Fig. 16. Sting of a scorpion.

that the sound acts as a warning to possible predators that there is danger in attacking the purring and venomous scorpion.

Male and female scorpions do not differ much in appearance. When the sexes meet, there is courtship, comparable to that of the false-scorpions described in an earlier chapter. Courtship begins by the male seizing the palpi of the female in his own, and, thus joined, the two walk to and fro, their tails upturned and intertwined, for an hour or more. At intervals the chelicerae of the two animals are brought close together, and again at intervals there is a remarkable swaying of their bodies, backwards and forwards. At last the male secretes a spermatophore and this is taken into the female body.

The embryos develop in separate outgrowths of the ovary, the far end filled with a nutritive fluid, which passes through a tube into the mouth of the unborn scorpion. The young ones appear not as eggs but enclosed in a membranous envelope, which is torn open by the sting. They then climb, slowly and laboriously, on to their mother's back, a peculiar form of motherhood, which they share with wolf spiders. Here they remain for a week or two until they moult for the first time.

They are colourless, less than half an inch long, rather weak, and liable to fall off their perch. At the end of their feet there is a special pad, which helps them to climb up again. Seven or eight moults bring them to maturity.

OPPORTUNITIES Not many British readers will have an opportunity of keeping and observing scorpions, but all will have realized that they are as interesting arachnids as any of their

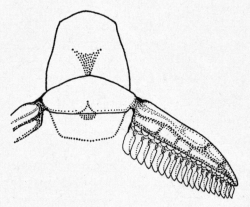

Fig. 17. Pectines of a scorpion.

class. In addition to their rather puzzling habit of stridulation, they carry on the lower side of the abdomen a pair of mysterious organs called pectines. They resemble combs and many ideas have been put forward to explain their function. Their use might well be examined by those who have the chance, for any new suggestions would be welcomed.

A recently described feature of the scorpion is its fluorescence. There is a substance in the exoskeleton which glows under the influence of ultra-violet light, and scorpion hunters therefore walk out at night, carrying an ultra-violet lamp, in the light of which the scorpions glow with a greenish-yellow fluorescence. British arachnologists will not experience this, but I mention the phenomenon because it recalls a letter which I received many years ago from a correspondent who reported that a spider crawling on the ceiling was seen to be glowing. I was asked for an explanation, which I could not give, but it seems that here is a chance for anyone to try an interesting experiment. There is little doubt in my mind that the spider was one of the genus Tegenaria, so common in houses in this country, and it seems possible that it had been in contact with some source of artifical light or perhaps only the sunshine. In any case, assuming that the original observation was sound, here is something that must surely suggest an investigation to any of my readers.

7

Other Orders of Arachnida

The inclusion of a chapter on scorpions, which do not occur in Britain, but which may be encountered by any who travel to the warmer parts of the earth, was justified because it should also encourage readers to take a wider view of the class, and to realize that as well as spiders, there are other orders well worth attention. In the present chapter this broadening of the horizon is carried a little way, for no one can justifiably think of himself as an arachnologist if he has no knowledge of the less popularized orders.

ACARI

The first to be mentioned is that of the mites and ticks, the order Acari, the relegation of which to this secondary position may surprise some enthusiasts. By long-standing tradition, not easily explained, there has been a clear separation between those who have become enthusiasts about 'acarology' and have in consequence neglected the spiders and scorpions, and those who have studied any or all of the other orders, while constantly declining to worry themselves about mites.

The reason for this separation must surely be sought among the mites themselves; and it cannot be denied that they differ in so many respects from the rest that the practice of those zoologists who put them in a separate class may be sympathetically accepted.

The number of known species of mites is great, running into several thousands and increasing rapidly year by year. It is probable that their total will ultimately reach the tens of thousands attained by spiders. Within this multitude their modes of life cover a wider range than that found in any order of the typical Arachnida. There are mites that are vegetarians, mites that are wholly parasitic; there are mites that live in fresh water and mites that inhabit salt water. Many of the parasitic mites are vectors of other organisms that bring diseases to the hosts on which they feed: in consequence mites and ticks have an importance in the field of economic zoology that no ordinary arachnid can share.

Mites in general are smaller than is the average arachnid. Their bodies show considerable modification of the simple division into two parts that is an arachnid characteristic, and many of them hatch as nymphs with six legs. Perhaps it is the fact that the adults possess eight legs that brought them into the arachnid world in the first instance, and gave them a precarious foothold in a province to which they had but an imperfect claim.

In this they cannot but recall the similar position of that fascinating group of marine arthropoda, the sea-spiders or pycnogonida. They, too, have in general eight legs in the adult and six in the larva, and the magical number 'eight' has given them an uneasy relationship to the Arachnida for nearly two hundred years. But some 'pycnogs' have ten legs and some have twelve, and if these numbers had been the prevalent counts, and eight had been the rarer, the taxonomy of the groups would have been different. Thus it has come about that the great names of those who have successfully studied spiders and scorpions are not the names of those who have with equal success studied the mites and ticks.

SOLIFUGAE

The second order to deserve mention in this chapter is the order of Solifugae, arachnids that are sufficiently numerous and con-

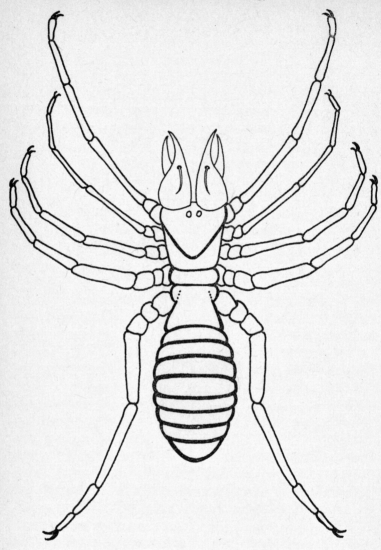

Fig. 18. A solifuge.

spicuous to have received several common names. They have been called wind-scorpions, camel-spiders and sun-spiders. They are to be found in the south of Spain, in almost the whole of Africa, Arabia, Iraq and India: in America they occur in a large area along the west coast of the States and on the north and west coasts of South America.

They have two outstanding claims to our notice, to which their attractiveness is largely due. The first is the amazing speed with which they can run: a correspondent once wrote to me, 'I have been made almost dizzy by their speed as they have raced round and round the inside of the upper part of my tent.' The second is the relatively enormous size of their jaws (chelicerae) and the ferocity with which they use these weapons when they are roused. In some species the chelicerae are half as long as the rest of the body, a proportion that justifies the description of the 'Solifugae as the most formidably armed animals in the world. Their bite is not venomous, but relies for its effect on the muscular strength of the jaws. So fiercely do they function that when a solifuge is consuming a beetle or a cockroach the crunching can be heard across the room.

Solifugae are often found in desert regions, where they protect themselves from the heat of the sun by digging burrows in which to rest during the day. They have an extraordinarily sensitive supply of long setae on their limbs, organs of touch which inform them of happenings in their neighbourhood. It is said that if one of these long setae, which may be 2 cm. in length, is just touched at the tip with a fine hair, the animal at once responds. Unlike many tropical Arachnida, they do not seem to adapt themselves very easily to life in Britain. Live African specimens sent to me by a former pupil survived the journey by air, but were somnolent and unresponsive to the care I gave them. Nevertheless, they are interesting animals and no opportunity to study them should be neglected.

RICINULEI

The pure romance of their history justifies a paragraph on the apparently rare order of Ricinulei. The first member of the order to be discovered was a fossil, and was described as an extinct beetle; this opinion was revised when a live specimen was found in 1838 in New Guinea. Further knowledge of these unusual creatures was acquired very slowly and when, in 1904, Hansen and Sorensen wrote the first collective account of them, it was based on eight species, represented by only twenty specimens.

Some really remarkable features characterize these arachnids. To the front edge of the carapace is hinged a cucullus or hood, which covers the jaws. There are no eyes. The cephalothorax and

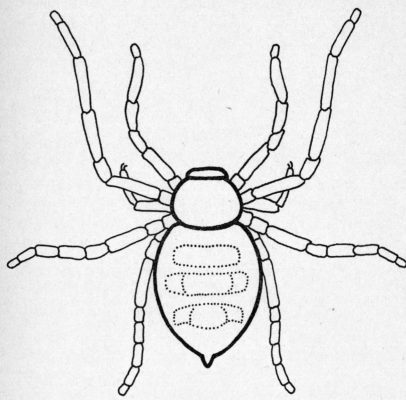

Fig. 19. A Ricinulid.

abdomen are held together by a hooking device, which must be unhooked to allow mating or egg-laying. Only one egg is laid. The males carry their sex organs on the tarsi of their third legs. It would be difficult to find a more remarkable collection of surprising features in any single kind of animal.

Ricinulei have been found in a number of scattered spots in tropical Africa and America. Little is known of their habits. Their reputation for rarity, which for years made the finding of each specimen something of a zoological triumph, has been modified in recent years, and in the special places in which they hide they are often numerous. In 1933 Ivan Sanderson brought home over three hundred specimens from the Cameroons; in 1964 another pupil of mine wrote from Sierra Leone 'I can get any number whenever I want them'; and in 1972 Dr R. W. Mitchell found that in the Mexican caves Ricinulei could be found literally by the hundred. Hopes exist for a further knowledge of one of the most remarkable orders of animals in the world.

Some early writer once described the Ricinulei as a 'primitive order', and for a very long time the adjective was unhesitatingly applied to them. More critical examination of their structure has shown that they are actually among the most highly specialized of all Arachnida, a fact which leads us to consider, very briefly, an order that is really primitive.

PALPIGRADI

The romance of the evolution of the Arachnida is, in its way, no less intriguing than that of the Ricinulei. Great difficulties have attended the unravelling of the story of the past, chiefly because of the imperfection of the fossil record. The order about to be discussed is of interest because of the light it throws on really primitive arachnids.

In 1885 Prof. Grassi described a small animal which he had found at Calabria in Italy, and realized that it belonged to a previously unknown order. About ten species are now known, mostly from warm countries.

The Palpigradi have certain specialized features, such as a prominent rostrum with the mouth at its tip, and their habit of walking with their first legs stretched forward in front, but their importance lies in their retaining a larger number of primitive characteristics than are found together in any other arachnid order. Thus they have a divided carapace in three pieces, eleven clearly defined segments in the abdomen and a sternum in four pieces. They have simple, leg-like pedipalpi and retain a pygidium in the form of a jointed telson. They have no eyes and no respiratory

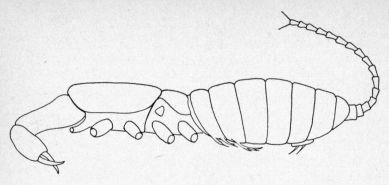

Fig. 20. A Palpigrade.

organs, they lead obscure, cryptozoic lives, and thus illustrate the principle that while a vertebrate palaeontologist looks for fossils an invertebrate palaeontologist seeks cryptozoa.

There is some evidence that they feed on the eggs of other small animals in their environment, a significant habit, perhaps suggesting that the earliest arachnids to arrive on land were able to survive because of a similar ability to digest whatever organic matter might be available.

A further point is the ability of present-day Palpigradi to survive the discomfort that accidental transport must involve; for they have established themselves in the Museum de Paris and have also made the journey to the slopes of Mount Osmund, Adelaide.

To any zoologist who happens to live where Palpigradi are available and who is willing to do work that is slightly off the beaten track, an intensive study of the biology of this order would be richly repaid.

8

Arachnology in the Field

Like every other branch of natural history, arachnology depends entirely on the obtaining of living specimens from their natural habitats and haunts; the dried or preserved corpses that occupy us in our laboratories can give us no more than a fragmentary picture of the vital whole. Arachnida must therefore be collected. The collection is seldom difficult, but it is more valuable and more interesting if it is carried on in an organized, intelligent manner, to be described below.

The beginner, however, in the first moments of enthusiasm, may start operations indoors. If his house contains any rooms such as attics, which are not in constant use, or if there are cellars in the basement, his chance of finding his first specimens close at hand are very good. There will almost certainly be webs in the corners, webs in the window frames, webs in the key-holes, with spiders likely to be living in them. At night there may be others parading the walls or ceilings—consider how often these wanderers have trapped themselves in bath or basin.

Unluckily, our houses today are not the homes of as many spiders as they used to be. There is a car in the garage instead of a horse in the stables, so that fewer flies are bred; and in many homes a fly is seldom seen. The vacuum cleaner and more especially the dried air of central heating also tend to reduce the numbers of insects on which resident spiders may feed. Nevertheless, the house is not to be neglected. In 1971 C. Tipton published

(B.A.S. Newsletter, No. 2) a list of ten arachnids he had found in and around a terrace house in the middle of Liverpool. Earlier than this, two of Dr. Bristowe's additions to the British List, *Oonops domesticus* and *Physocyclus simoni*, were found indoors.

For the earliest expeditions into the countryside the simple methods often described can be hopefully employed. These are (i) the beating and shaking of the hedges and lower branches of trees over a newspaper or inverted umbrella, (ii) the sweeping of long grass with a stout canvas net, and (iii) the sifting of fallen leaves over a newspaper. All these methods will yield different kinds of spiders and harvestmen, with an occasional false-scorpion from the sieve, and these will provide a sound introduction to the spiders of the neighbourhood. Pursued without further elaborations, these methods will produce about a hundred species during the first year.

To continue in this rather vague manner soon becomes un-rewarding, and in consequence so tedious that interest flags. To acquire a really worth-while knowledge of the arachnids of one's district, and to support that knowledge with a good, representative collection, will take time, and the intelligence, foresight and planning that are desirable will effectively prevent field-work from becoming a bore. The time aspect may well be mentioned, for the longer one has been a collector the greater is the surprise and pleasure at turning up a species that one has not met before. My own experience provides an example. I began the habit of making fairly frequent visits to a wide ditch near a wood in Herefordshire about 1924. In 1950 I found for the first time two specimens of the comparatively rare harvestman *Anelasmocephalus*.

What may be called the more advanced type of collecting takes note of the fact that every species of arachnid tends to occupy one particular micro-habitat. Just as you would not expect to find a cormorant on the slopes of Snowdon, so you will not find in a damp, dark cave the jumping spiders that are active in a sunny woodland glade, and so on. In consequence, the different places in which spiders and other arachnids may be sought with expectation of success demand of the collector that he takes different tools on his searchings. The stick and the morning paper are correct for the hedges and trees, on another day the sweeping net, and on another the sieve. Further, a day may be given to lifting loose bark from trees, especially fallen ones, for which operation a stout knife or an old chisel should be carried. For the water spider and the pirate spiders a net is needed; for the many that live among and below the grass roots a trowel. The lesson from this paragraph is that each day's outing should have its special aim, not a vague one.

To this may be added the observation that most of the habitats open to arachnids may be occupied by different species at different seasons of the year. Many spiders are seasonal migrants, moving from the ground to the shrubs as they grow, as does Agelena, or from the trees to the ground, as do some Clubiona. Recognition of this fact may help to teach much about the actual life-histories of the species concerned.

When the beginner in arachnology has sufficiently followed these preliminary stages, he has come to know most of the commoner species that he is likely to find, and he has acquired some knowledge of special habitats. He has now reached the point from which he should embark on one of the most rewarding sides of field work, the study of ecology as shown by Arachnida. Henceforward he will set out suitably equipped and mentally prepared to seek the answers to one or more of four basic questions:—

1 Which species occur in any one habitat?
2 How many such species are there in it?
3 How many specimens of each species are present?
4 How do their numbers vary throughout the year?

Other problems will turn up in later years, but the above indicate the first aims of the ecologist; and one of the real merits of arachnology is the fact that it is so well suited to ecological investigation.

Hence the eco-arachnologist divides the scene of his operations into the three zones or layers that have long been recognized by botanists:—

1 The Ground Layer, up to six inches above the level of the soil and as far below it as is profitable.

2   The Field or Shrub Layer, which may be taken to extend
about a yard above the ground.

3   The Tree Layer or 'Canopy', which includes all habitats
above the shrubs.

Each of these layers is attacked by methods best suited or even
peculiar to itself and it is obvious that the three simple elementary
methods already mentioned, namely, sifting, sweeping and beating,
are the ways in which to begin at ground, shrub and tree layers
respectively. Devices additional to these may now be considered.

On the ground hand-picking or stone-turning are not to be
neglected. They are restful, if crude, methods suitable for the
aged or for the pleasant moments of the coffee-break or tea-time
picnic. It should not be forgotten that some of the recent additions
to the British list have been found in just this simple way.

Sifting, however, covers the area more effectively, and a word
may be said about the sieve. The ordinary circular sieve is awk-
ward to carry about and is even worse on a bicycle. I have used
a boat-shaped sieve, made by folding in half a piece of wire-
netting and joining the sides. It can be flattened for transport
and opened for use. Mr. Croker has described, in Vol. 1 of the
Bulletin of the British Arachnological Society, an ingenious form
of home-made collapsible sieve, which any enthusiast can construct
for himself.

A difficulty that often attends the appearance of a desirable
specimen on the ground or the sheet is that of actually picking it
up. Sometimes the creature in question may be guided or coaxed
into an empty tube, but a piece of apparatus that is popular with
many naturalists enjoys the delightful name of 'pooter' (Fig. 21).
This can be easily constructed in many and various forms, all of
which are used in the same way—to suck the animal into the wider
tube, where a piece of gauze or muslin prevents it from continuing
its journey into the lungs.

Probably the most widely used method of extracting small
animals from leaf litter and vegetable debris is the use of the
Berlese Funnel (Fig. 22). This is especially valuable in the search
for false-scorpions, which cannot be sifted from leaves whenever
the leaves are damp.

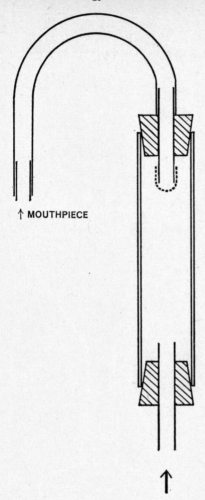

↑ MOUTHPIECE

Fig. 21. One form of pooter.

A Berlese Funnel consists essentially of a containing reservoir, inside which is a horizontal sieve. At the top there is an electric light bulb and at the bottom a funnel-shaped exit leads to a bottle of preservative. The matter in which the cryptozoic animals are living is put into the cylinder and rests on the sieve. The light is switched on. The animals tend to move away from the heat

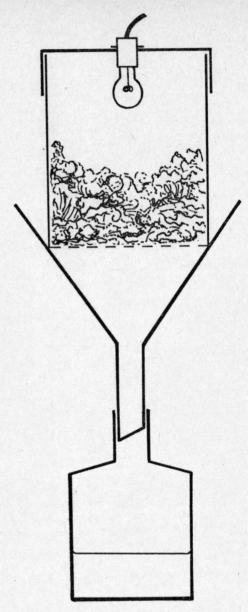

Fig. 22. A Berlese funnel.

and from the light, so that they burrow lower and lower until they fall through the sieve.

A modification of the Berlese Funnel is known as the Tullgren Funnel. In this the warmth is supplied by a stream of hot water circulating in a tube wound spirally round the cylinder. It is obviously only possible to set up such a device in a properly equipped laboratory, whereas the type with its electric bulb can do its work in one's bedroom while one restfully sleeps. Another alternative consists in the placing of a moth-ball on top of the heap of vegetation. As the naphthalene slowly volatilizes its heavy vapour diffuses downwards into the vegetation and repels the inhabitants towards the sieve. The method is not quite fool-proof for there is a risk that too much vapour may be formed and kill the animals instead of merely moving them on.

The construction of a Berlese Funnel is simple. The size of the cylinder depends on the scale on which the arachnologist is working and may be no larger than a domestic coffee tin, 10 cm. or so in diameter although, in the laboratory, a larger size is convenient. An ordinary oil tin may be adapted, and has the advantage that its top is already funnel-shaped. It is turned upside down and the bottom cut off. The best results are obtained from the funnel if the lamp is no stronger than 25 watts and the extraction is allowed to go on for twelve hours. The use of any funnel method has one great advantage. Quantities of the vegetable debris that is to be treated can be put into sacks or plastic bags, taken home and searched at leisure: a delay of a few days matters not at all and the convenience is obvious.

There is, of course, much to be said for any method that causes animals to catch themselves, and thereby save the zoologist the labours of the chase. Entomologists discovered this years ago, when they painted the bark of trees with patches of 'sugar'; the arachnologist's parallel to this is the use of the pitfall trap.

The object most often used for this is a one-pound jam jar or a plastic cup, which is sunk in the ground so that its rim is flush with the surface. The theory is that the wandering arachnid falls into this and cannot climb out, and the trap is visited at convenient intervals of time, when its contents are extracted.

This extreme simplicity needs some elaboration. Glass jam jars have the advantage that their shape and their smooth surface make escape all but impossible; plastic cups, however, can be provided with drainage holes in the side, permitting the exit of rain water if the soil is porous. Traps intended to supply live animals should be visited daily and may well be provided with a few stones or leaves under which the animals can shelter. Traps that are visited once a week should contain some preservative, such as 2% formalin.

The siting of a trap is important. If it is in an exposed position it may suffer from passing sheep, cattle, dogs etc; if it is in an area of thick vegetation few arachnids are likely to make their way into it. The risk of rain is always present, and the covering by some sort of roof is a good addition. Experience in the area under investigation is by far the best way of deciding on the form of trap to be used and where to site it.

Then the tendency of arachnids, like most other small animals, to seek and come to rest in the shelter of some solid object may be exploited. Crumped pieces of paper concealed in dry ditches and at the base of hedges may often be found, when re-visited and straightened, to have been adopted by passing arachnids as places of shelter. In the same way, stones provide natural sanctuaries for many, and these can be supplied with all that they want in the form of tiles or pieces of slate, suitably placed. These seldom fail to be found to be covering a spider or two when they are lifted a few days later.

Rising now to the Field or Shrub Layer, we remember that the simple method was the use of the sweeping net. This is an essentially crude way in which to go to work, and is rather likely to injure the captives. It is, however, the easiest and most convenient way by which to determine the inhabitants of long grass or rough herbage. It is a method that is also favoured by some arachnologists who wish to make a quantitative study of some particular location. They make x sweeps, each of y feet in extent, at intervals of z yards, and from a count of the produce they calculate the figure they want.

Beating is naturally the best way of dislodging arachnids from hedges or from creepers on walls and houses. The falling animals are arrested on a sheet of paper, a tray or an umbrella. I used to have some success with a triangular object ('spiangle', Fig. 23), which is easily made with a wooden frame and a canvas base. The point is thrust into the bush and the leaves above it are shaken.

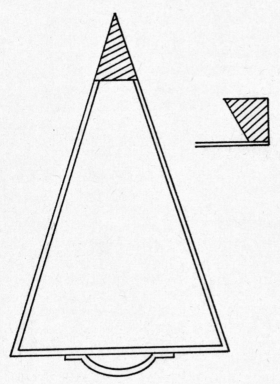

Fig. 23. A 'Spiangle'.

A feature that is a help is the shape of the wood, a section of which is shown. The arrivals on the sheet run to the sides and, meeting the overhanging rim, will usually crouch there instead of running over the top. The device is more handy and efficient than an umbrella.

A characteristic of the Shrub Layer is that many of the webs of web-spinners are to be found in this zone. There is therefore something to be said for a quieter, if slower, inspection of a hedge or bush for the presence of webs. These are more easily seen when the sun is behind them and low in the sky. The form of the web often indicates the nature of its maker, who may be tempted out of her hiding place by touching the threads of her web with the prong of a vibrating tuning fork.

Finally, the Canopy, represented for arachnologists by the lower branches of trees, is also a place where beating is useful. There are, however, several species that live on the bark of trees and hide themselves in its crevices. They are not always easily found, and experts recommend a brushing of the bark with a soft paint brush.

We, who have been accustomed to putting grease-bands round our fruit trees, know that spiders sometimes fall victim to them, and that they are in poor condition when they have been extracted from the gum. But a comparable method is the tying of bands of corrugated cardboard round trunk or branch. These provide artificial shelters, which are attractive to tree-spiders, just as tiles and slates are useful on the ground.

When an arachnologist has mastered and proved all the devices and dodges briefly outlined above, he will find sufficient opportunity for extending his work to further fields. There are still rewards awaiting the observant or the fortunate in the areas made famous many years ago by the collectors of butterflies: the New Forest, Burnham Beeches, the Norfolk Broadland, Box Hill, the Cambridge Fens. Add to these the summits of the mountains, the depths of caves, the crypts of cathedrals, as well as chalk pits, rubbish dumps and slate quarries, and the arachnologist finds that he is never without promising objectives.

An important extension of the collecting of arachnids in the field is the raising of it above the level of a mere accumulation of species. For example, the name of the tree, shrub or plant on which any unusual species has been found, should be noted and recorded. The date, or at least the season, is also valuable. Ecology has developed its quantitative side, as witness Bristowe's estimate of two million spiders per acre on a Sussex field; and most recent ecologists record their findings with the help of graphs and histograms. All those who are seriously interested in field-work should study closely a paper by E. Duffey, *Ecological Survey and the Arachnologist*, in Volume 2 of the British Arachnological Society's Bulletin. The work of an expert, it is also the best possible proof of the fact that, though science should be learnt by practice, the scientific attitude may be inculcated and encouraged by reading.

9

Arachnology in the Laboratory

Returning from his out-of-doors activities, and carrying the spoils that skill, experience and a slice of luck have provided, the arachnologist proceeds to the next stage, the work in the laboratory.

A word about the laboratory itself. The essentials of an arachnologist's home laboratory are not elaborate. A start is made with a table, ideally with a formica surface, placed in front of a window, ideally facing the north: add to this a cupboard for storage and shelves for the collection of bottles. The really important item is a microscope, with its accessories. Although a microscope cannot be too good, many of the instruments intended for schools, and produced at moderate prices, are efficient enough for all ordinary arachnid work.

With this modest equipment the first task may begin, the naming of the specimens captured before they are preserved in the Collection. Their bodies are best examined lying in spirit in a white saucer, first under a hand lens, then under the low power of the microscope, and illuminated from above. When their limbs have been spread out their characteristics are compared with the printed descriptions: as is well known, a dichotomic table or 'key' is the most practically helpful form in which the details to be examined may be set out.

For harvestmen and false-scorpions the difficulties are seldom great, and a little experience soon makes the identification easier. The best guides are the *Synopses of the British Fauna*, published by the Linnean Society and intended for just this purpose (see Chapter 10).

Spiders present a more difficult problem. Although there are many more of them, the determination of the family of a given specimen is seldom in doubt, and several books are available for help with this first step. Very soon the family to which a British spider belongs can be seen at a glance: for example their long spinnerets betray the Agelenidae, as do the huge eyes of the Salticidae, the sideways legs the Thomisidae, the stout spinnerets the Gnaphosidae, the orb-web the Argiopidae, while most of the very small spiders belong to the Linyphiidae.

The determination of the species is not so easy. Apart from perhaps a score or so in which the colour or the pattern or some other characteristic gives helpful evidence, the fact is that the use of the microscope to examine the genitals is essential. From this it follows that immature specimens are often indeterminable, and can only be ascribed to the comforting category of 'spec. not det.' or 'spec. juv.'

For the rest, the arachnologist appeals to the splendid work of Locket and Millidge, gratefully thanking Fate that these volumes were printed before he took to the study of our native spiders. In their pages there is to be found a description of every species so far recorded, together with illustrations of the female epigyne and the male palp. Female specimens are therefore placed with the ventral surface uppermost and examined through the microscope, and the epigyne compared with the drawings of the species in the appropriate family or genus. For males the left palp, which may have to be removed for the purpose, is examined from the inner side. The separated palp can either be included in the tube with the rest of the body or mounted as a microscope slide.

During this process the spider has been immersed in alcohol. Sometimes the worker may think that his task would be easier if the specimen were allowed to dry. There is no harm in this if the drying does not last too long, and the body will soon reabsorb

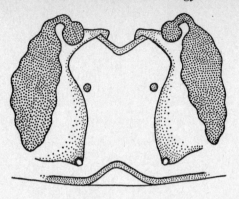

Fig. 24. Epigyne of *Tegenaria saeva*.

fluid when returned to the tube. If by misfortune any specimen has become absolutely dry and rigid it may be regarded as ruined. In a case of exceptional value an attempt at salvation may be made by the method suggested by Van Cleave. The desiccated body is dropped into a .5% solution of tri-sodium phosphate and left there for at least twenty-four hours. It may then be transferred to alcohol and kept—for what it is worth.

The simplest method of keeping a collection of spiders has been determined over a number of years. It depends on the fact that the soft bodies of Arachnida will shrivel and wrinkle if allowed to dry; therefore they must be kept in tubes of 70% alcohol to which a little glycerine has been added. This preserves the suppleness of their bodies. The tubes are either corked or plugged with cotton wool and inverted in more alcohol in a stoppered bottle. Bottles with ground glass stoppers are expensive, and many screw-topped bottles with plastic tops are almost, but not quite, as efficient.

The perpetually serious disadvantage of any collection of spirit-specimens is the evaporation of the alcohol. Just as tiny particles of dust find their way into a microscope, however carefully it is treated, so alcohol will invariably evaporate, even though slowly, from any collection of arachnids. The glycerine, which in any case makes the specimen rather messy, will now allow the growth of fungal spores, and the specimen is lost.

For many years all zoologists who have to deal with spirit specimens have wished for a substitute for alcohol, and some have recommended iso-propyl alcohol, which evaporates more slowly. In 1956 Owen and Steedman introduced the use of propylene phenoxytol. It is used for the preservation of spiders, etc. that have previously been fixed in alcohol at no greater strength than a 2% solution. Cooke recommends the use of this solution for killing arachnids as soon as they are caught, for they die in it very quickly, with their limbs extended. They are then left in 70% alcohol for fixing and finally stored in the propylene phenoxytol.

A collection is to be arranged according to the ideals of the collector. If these are put as high as possible, there will be a separate bottle for each species, and the tubes therein will contain the captures of each locality and of each year, separated from one another. Less ambitious collectors will give a bottle to each genus or even to each family.

Each tube should contain a label on which is written in pencil the name, date and locality of the contents. Two kinds of tubes of special merit may be distinguishable, namely, those that contain 'authoritatively named specimens' and those that contain 'figured specimens'. The former are probably less numerous than they were in our collections of fifty years ago, when the literature needed to establish the names of our captures was non-existent or inaccessible. We were then compelled to send our difficulties to wiser and more experienced colleagues, who, with the helpfulness that is characteristic of arachnologists, repressed their inward irritation, and kindly returned them, duly christened. These specimens therefore had an enhanced value, because their identities were not in doubt. It was a good plan to write their names and that of the authority on pink paper.

Published work is usually accompanied by illustrations, and the material from which these have been drawn also acquires an added interest. If the source of the drawing has been mounted as a microscope slide, an extra label should be stuck on it, bearing the words 'Figured Specimen' and a reference to the publication concerned. If it is contained in a specimen tube, the same information may be written on pale blue paper.

A laboratory, however, is more than a storeroom; it is, by tradition, a work-room where anatomy is studied by dissection. It is by the dissection of a few examples that knowledge of the structure of an arachnid is obtained, and the printed descriptions given a better sense of reality.

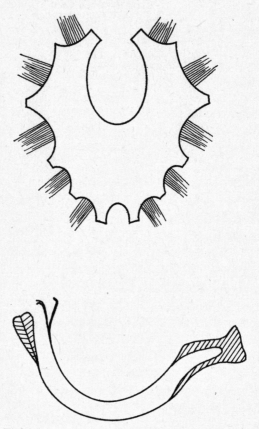

Fig. 25. Endosternite and oesophagus of a spider.

A large spider, such as *Tegenaria saeva*, is no harder to dissect than an earthworm or a cockroach. Embedded in wax in the usual way, it is opened dorsally, using a sharp and very small blade. The heart, the alimentary diverticula (the so-called liver) the silk glands and the tracheae are easily found in the abdomen. The carapace can be lifted off, revealing the poison glands, the endosternite and the characteristically curved oesophagus. The first dissection is likely to be disappointing, because of the strangeness of arachnid architecture; obviously it should be repeated and followed by the dissection of a harvestman of the species *Phalangium opilio*.

Many of the internal organs may be removed, stained and mounted as microscope slides without any special technique. A microscope is invaluable for many purposes. The setae and spines, with which the legs are covered, are of various forms and dimensions. If a spider's leg is over-boiled in dilute potash they are freely shed, can be collected in the jet of a pipette and transferred to a slide where they may be allowed to dry, and then covered with a drop of adhesive and a cover-slip.

The naked leg, if picked up, dehydrated and mounted, will now give a clear outward view of the slit-sense organs or lyriform organs and, on the tarsus of the fourth leg, can be seen the tarsal organ, an organ of chemotactic value.

Some naturalists like to boil an entire spider in potash until only the chitinous exoskeleton remains, and then mount the whole, flattened, under a large cover-slip. The process is not a very intellectual one, but if the specimen is carefully arranged it forms a preparation in which nearly all the characteristics of an arachnid can be seen. It cannot compare with the pleasure of making a delicate dissection and successfully extracting and mounting the pertinent parts.

In any thorough examination of a spider, the sex organs demand special emphasis. The epigyne of the female may well be completely removed from the lower surface of the abdomen and gently warmed in dilute caustic potash to remove the soft tissues. It may then be mounted as a permanent microscope slide, a process which for some species requires much care in getting the object to lie in the best position for direct viewing. The traditional canada balsam may be used, but a strongly recommended mount is a synthetic resin, dimethyl hydantoin formaldehyde (DMHF). Alternatively, the genitalia may be kept in a very small glass tube included in the larger tube with the rest of the spider.

The male palp is easier to deal with, and perhaps more interesting. In life it is ordinarily in a contracted condition, the form in which it is examined when trying to name the species. But when used in sperm-transfer it expands into a complicated organ, the details of which need the microscope for proper examination.

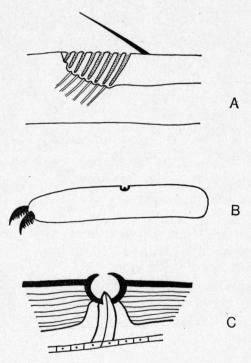

Fig. 26. Lyriform and tarsal organs on a spider's leg with section through latter.

A. Lyriform organ.   B. Tarsus, with tarsal organ.   C. Section through tarsal organ.

Expansion may be brought about by gently warming the palp in dilute caustic potash, and in this state it may be scrutinized as it lies in spirit in a watch glass. Opinions differ as to the next step. Some arachnologists now mount the expanded palp in the ordinary way; others object that in this fixed position it cannot be satisfactorily examined and prefer to keep it separately, like the epigyne, in a very small tube.

Harvestmen, even the largest species, are less easy to dissect than spiders and less rewarding to one's attempts to do so. But there are two compensations. As mentioned above, harvestmen are unique among Arachnida in the possession of an intromittent organ, concealed inside the abdomen, and this organ is different

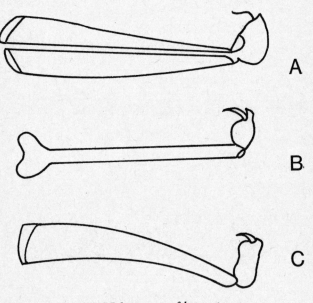

Fig. 27. Male organs of harvestmen.

A. Mitopus morio.　B. Oligolophus agrestis.　C. Oligolophus spinosus.

in appearance in different species. It may therefore be extracted and mounted as a permanent preparation. The extraction is not difficult, for the penis is tough and chitinous: it can easily be picked out from the surrounding soft tissues, cleaned, dehydrated and mounted.

It is also interesting to make a closer examination of the ventral sternites than the usual diagrams display. The same animal may be used after the penis has been extracted. If the four legs of one side are removed and those on the other side are pulled outwards, the cephalothorax opens, revealing the actual positions of the ventral plates. If the genital operculum is pulled backwards the vestigial first abdominal sternite is exposed (Fig. 28).

False-scorpions are too small for ordinary dissection, but the student should not fail to practise Chamberlin's method of so treating a specimen that all its characteristic features can be made permanently visible. His prescription is as follows:—The chelicerae and pedipalpi and the first and fourth legs of one side are removed from the body, dehydrated, cleaned and mounted. If one of the chelicerae and one of the pedipalpi is allowed to dry, their fingers

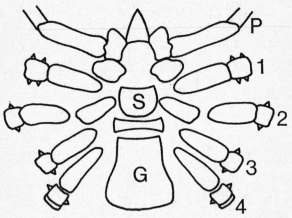

Fig. 28. Ventral view of harvestman's body.

P. Pedipalp.   S. Sternum.   G. Genital operculum.   1–4. Legs 1 to 4.

will open and in this state may be mounted. The rest of the body is gently warmed in caustic potash, washed, stained with magenta or fuchsin, cleaned and mounted. One or two slides will accommodate all the small cover-slips necessary.

The cutting of sections with a microtome is the usual way in which the internal organs of a small animal are studied, but it is a difficult process when spiders are the animals being studied. The reasons for this are the toughness of the chitinous exoskeleton and the hardness of the ovaries after they have been fixed. The former difficulty may be avoided by using a spider just after it has moulted and when it is still soft, but the chances of coming upon such a specimen are rare. The latter may be met by removing the ovaries and treating them separately.

Those who have had some experience in the use of the microtome may care to try the procedure successfully used by Millot in 1926. The dehydrated body is transferred from alcohol to a mixture of ether and alcohol for twenty-four hours, thence for the same period to celloidine solution, and after washing in two changes of toluene is embedded in paraffin wax.

Alternatively a more recent process due to Hopfmann consists of three stages:—fix in Carnoy's fluid, pass through methyl benzoate, methyl benzoate with celloidine, benzene, and embed.

The sectioning of harvestmen is easier, and the normal less complicated process may be followed. A Sixth Form pupil of mine successfully cut some excellent series of sections of Phalangium in the school laboratory.

An arachnological laboratory should also house some living specimens, so that their life-histories, habits and behaviour may become known to the arachnologist at first hand. The possibilities in this side of arachnology have been outlined in Chapters 3—5, and the following remarks may be added.

The popularity of bird-watching and more recently of badger watching may incline the arachnologist to think that spider-watching should be an outdoor occupation. Emphatically this is rarely so. Efficient observation of the behaviour of small animals demands continuous concentration and the making of a real effort; out of doors there are too many distractions and it is seldom possible to settle down in comfort. Moreover, arachnids are so small that one must either get close to them or make use of a lens, and the austerity of a laboratory is a better environment for this than any open-air situation. There is ample scope for original work on invertebrate behaviour, which does not get its fair share of attention. Witness the journal *Behaviour*, Volumes 43 and 44 of which contained seventeen contributions, all but one of which were concerned with vertebrates.

Laboratory animals need to be fed. Spiders may be supplied with live flies, when these are available, or with smaller spiders, or with almost any insect that is not distasteful. Harvestmen need

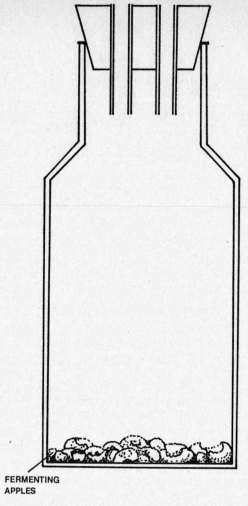

FERMENTING
APPLES

Fig. 29. Bottle-trap for *Drosophila*.

only bread, fat and water, since they will eat dead matter. False-scorpions will take springtails eagerly, and will accept most of the other cryptozoa that are sifted with them from litter.

Many captive arachnids have been brought up on *Drosophila*, the ease of breeding which is well known. To secure *Drosophila*, pieces of chopped apple are kept warm for several days, that fermentation may set in; this is then put in a bottle-trap, shown at Fig. 29, and left in the open in warm weather. Attracted *Drosophila* can enter but cannot escape from the bottle, which is taken home and the flies removed with a pooter. They can be fed directly to the arachnids or used as the parents of a permanent culture. For practical details of breeding generation after generation of these insects, the reader should consult the specialist publications, such as B. Shorrock's: *Drosophila* (Ginn, 1972).

There are many zoologists who wish to record their observations by photography. There is a difficulty in the photography of spiders which is to be seen in many of the attempts to photograph them, the fact that a close approach of a camera to a small creature results in a picture with part of the subject sharply focused and a part, sometimes not a very small part, blurred. Such imperfections are more serious when the spider is alive and not necessarily co-operative. These problems can only be met by the use of specialized apparatus. From the advice of an exceptionally successful photographer of spiders I quote the following:—'Photographic details. Exacton 11b 100 mm lens (50 mm Pancolor—2x converter) and 90 mm extension, f22. Twin electronic flash. Microneg Pan film, developed in Acutol 1:20 for 7 minutes.'

This chapter should make it undeniable that there is much arachnology to be done when work out of doors is impossible.

10

Arachnology in the Study

An arachnologist's work in his study is an important sequel to his experiences in the field and laboratory: it is the place where his ideas are preserved, his notes written up and developed, his correspondence conducted and his reprints stored.

The preservation of ideas is essential, for arachnology cannot be pursued for long before it creates ideas of the most diverse kinds in the mind of the arachnologist. Should it not do so the conclusion is inevitably that the student has found himself in the wrong box: he may then be advised to seek alternative occupation as a shop steward or as one of Her Majesty's Inspectors of Schools.

Those, however, who have avoided these delectable professions will, with equal inevitability, find that ideas come, and at the oddest of moments. They will come and whisper in our ears as we lie sleepily asoak in a hot bath, or as we mechanically hoe the weeds that grow between our cabbages, or even, as Cloudsley-Thompson has written, while at the sink we are helping with the washing-up. In fact, they delight in choosing moments when pen and paper cannot be found to impede their flight. As Ruskin said, "Passing thoughts are fleeting; down with them in the note-book; paper is cheap enough".

Paper is now expensive, and the value of our note-books has risen accordingly. There should be at least three of these.

The first is the *Laboratory Log*. This should contain a series of dated records of work done at the laboratory bench. There should be drawings and diagrams and memoranda of any difficulties or problems that will need further attention. Anything that is curious, unusual or surprising should find a place in it, for these will have a greater value in the future than they may seem to have at the moment. When in doubt about an entry, decide to include rather than to omit it.

The second is a *Scrap Book*, into which are pasted such newspaper cuttings and pictures that are concerned with any arachnid. Spiders and scorpions do not appear in the press as often as 'teenagers' and 'spokesmen', but on occasions they earn a paragraph. The scrap book will not be a large one, but it is a convenient form in which ephemera can be preserved.

The third book is the *Arachnological Journal*. This may take the form of a diary, and in it is written a record of expeditions, captures, and above all of periodic thoughts and the ideas already mentioned. This book is no less than a material form of the arachnologist's mind, expressing his wishes and hopes and recording his successes, his failures, his progress and his set-backs.

All these books should be of the loose-leaf type. Pages can then be rearranged, or taken out without loosening the other half of the leaf. Sections can be expanded, and lined paper can be mixed as required with blank paper or graph paper, so that each one acquires an individuality which gives it a real personal value.

These books should have at the beginning a few pages recording their contents. Two pages with five columns on each side give twenty columns, headed alphabetically, with IJ, PQ, UV and WXYZ as composites. In practice this is usually an adequate guide to the numbered pages that follow.

The arachnologist, if worthy of the name, is unlikely to follow his chosen line for long before the need to write an account of something that he has noticed becomes insistent. It is most desirable that he should write it well, for there is a tradition among literary critics that scientists do not write good English.

That there is a germ of truth in this is shown by the existence of several essays intended to teach the scientist how to prepare his contributions to learning and it is undeniable that the writings of many British scientists compare unfavourably with those of their French colleagues. The piece of advice that may be offered to an arachnologist is that he should cultivate the ability to write clear forceful prose.

This is a personal matter, which is quickly followed by the problem of publication. Scientific matters are normally printed in scientific journals, most of which are well-known, and many of which tend to specialize in the kind of matter that they accept. The most widely-opened source for a beginner is probably the magazine of his local Natural History Society. These journals differ from newspapers and ordinary magazines in that they seldom offer authors payment for their work; they rightly regard the privilege of appearance in their columns as a sufficient reward. Scientists accept this state of affairs.

An invaluable feature of scientific journals is that most of them print, usually on the inside of the back cover, the conditions which they ask their contributors to understand. These helpful remarks should always be digested by the writer before he begins the composition of his paper. It is not only a matter of using quarto paper instead of foolscap (or vice versa); it is important to know the most acceptable length of a paper, and whether limits are set at 1500 or 3000 or 6000 words: and to know whether the *Journal of the Linnean Society*, or the *Journal of Zoology*, or *Natural History Magazine*, or *Animals* is the most likely to accept the paper that he has to offer. Nearly always wisdom recommends a preliminary letter to the Editor, giving an outline of the proposed contribution and asking him whether he would be interested in seeing it. To do this may save time, expense and disappointment.

The matter of illustration may also be mentioned. Photographs call for little comment; they should, if possible, be of half-plate size and printed on glossy paper. Of line drawings and diagrams editors often say that they must be drawn in Indian ink on Bristol board. I always 'forget' this costly elaboration and do my diagrams on note paper with the pen I am using now. The editor then either uses them as they are or has them redone by a resident draughtsman. I am not recommending that you should follow my bad example.

In his study the arachnologist pursues the most important, the most exasperating and the most elusive task that confronts every scientist, whatever be his special field—the task of keeping abreast of advancing knowledge. Books like those mentioned in the next chapter are helpful when they appear, but books on Arachnida are not published very frequently, and when they reach the expectant purchaser the most recent paragraph in their pages will inevitably refer to some item of twelve or eighteen months age.

The latest news from the arachnological front is only to be found in the appropriate journals, produced all over the civilized world, often in publications of which the inexperienced arachnologist has never heard, and as often written in a language that he cannot read. A virtually complete list of those printed each year is published in the annual volumes of the *Zoological Record*. Because of its essential size this invaluable work is to be had in sections, dealing separately with different groups. The part 'Arachnida' (largely concerned with mites) has risen steadily in price during the past thirty years. The 1968 section, published in October, 1970, was £2.00; the next issue jumped to £6.50, a price that obviously puts it outside the 'income bracket' of all save tycoons and manual workers, who are seldom interested in spiders.

Fortunately, there are valuable alternatives, readily available for others.

The first of these is to become a member of the British Arachnological Society. This admirable body was founded in 1959, growing out of the British Spiders Study Group of an earlier date. Those interested should write for particulars to the Hon. Secretary, Mr J. R. Parker, at Peare Tree House, The Green, Blennerhasset, Carlisle, for details. They will learn, among other things, that the annual subscription is £2.00, that a Bulletin is issued four times a year, that field courses are arranged in different areas several times a year, and that there is an Annual General

Meeting, always well attended and always enjoyable and valuable. The *Bulletin* is, however, the most valuable side of the Society's activities, for it contains news of all sorts, articles about new discoveries and discussions of new ideas, and is, in fact, the ideal method for uniting the arachnologists of Britain and of simultaneously encouraging the progress of arachnology.

For more ambitious arachnologists another organization is well worth consideration. This is the Centre International de Documentation Arachnologique, the Sécrétaire-General of which is Prof. Max Vachon, at 61, rue de Buffon, Paris 5ème. The Society co-ordinates arachnology throughout the world and maintains a representative in most countries, publishes yearly a list of recent and forthcoming papers and books, and arranges an International Congress on Arachnology to be held once in three years in a different country. This Congress was held in 1971 at Brno and will meet in Amsterdam in 1974. The minimum subscription is at present 25 francs, which may be paid through Mr Parker of the B.A.S.

Older than either of these societies is Toa Kumo Grakkai, the Arachnological Society of East Asia, which was founded at Osaka, Japan, in April, 1936. Membership is limited to residents in east Asia, but arachnologists from other countries can become subscribers, at a cost of 1000 yen per copy, to the society's journal, *Acta Arachnologica*. This is produced twice a year and always contains valuable contributions, of which those written in Japanese are followed by an English summary. No serious arachnologist is likely to regret making contact with this Society. Its president is Dr Takeo Yaginuma, Biological Laboratory, Ohtemon-Gakuin University 230 Ai, Ibaraki, Osaka 567, Japan.

The youngest of the national arachnological societies is the American Arachnological Society, which, repeating the pattern of our British society, grew from a modest study group, the Arachnologists of the South West. This was formed by a group of enthusiasts in April, 1963, and its success under Dr. B. J. Kaston inspired the foundation of the American Arachnological Society in January 1973. There can be little doubt that our American colleagues will conspicuously succeed in their aims of promoting the study of Arachnida and in furthering co-operation between arachnologists all over the world. It also produces its own Journal, which is sent to all members. The Secretary and Treasurer, to whom the annual subscription of $10 should be sent, is Dr M. E. Thompson, 1000 N. Durfee Avenue, So El Monte, California 91733.

The receiving by authors of reprinted copies of their papers has been mentioned above, and almost as soon as the name of an arachnologist becomes known as that of a serious and competent contributor to our science he will begin to receive gifts of these, 'With the Compliments of the Author'.

First, attention may be drawn to the word 'gift'. Some journals present their authors with reprints up to a certain number, varying from a dozen to a hundred and fifty; some allow the purchase of further copies beyond their fixed number; some ask for payment for all copies supplied. Even free reprints cost the author the postage on every copy that he distributes, a deduction from which is that one should not omit to write a line of grateful acknowledgement, and if possible to return the compliment by sending a reprint of one's own.

An undeniable characteristic of arachnology is the generosity with which its followers interchange their 'Separates', and the number of these items that accumulate on the study desk. In the beginning, while they number but a dozen or so, there is no great difficulty in remembering their existence and recalling their contents. When they exceed twenty, this is not so easy, and well short of fifty it becomes impossible. Hence there arises the problem of storage and indexing.

Four methods of storage are possible.

1   To have them professionally bound at intervals. The disadvantages of this, the customary method in large libraries, is the expense, which is considerable, and the length of time during which the papers cannot be consulted because they are in the binder's hands.

2   To keep them in a set of box files. Almost as satisfactory a method, especially if one makes the containers onself.

3 To learn the art of bookbinding oneself, and bind the reprints as they accumulate. This system is much to be recommended.

4 To keep the booklets in files or folders. This is perfectly satisfactory if the collection is not too large. Portfolios are easily made at home, and their sizes are readily variable to fit their contents.

All volumes, files or folders must be labelled and numbered.

The best way of ensuring immediate tracing of any particular paper is the card index system. Here choice lies between the purchase of a set of cards with coloured margins already printed and contained in an impressive case, and the preparation of an equally efficient substitute of bisected post cards ($3\frac{1}{2}'' \times 2\frac{3}{4}''$), kept in any convenient box.

The cards themselves fall into three series. In one they are headed with the author's name; in the second with the branch of arachnology to which they refer; and in the third with the name of the order, family or genus mentioned. The making of this kind of card index is not difficult, and is an essential part of arachnology in the study.

Make no mistake, your collection of reprints will grow rapidly and its volume increase beyond your expectation. My own First Series amounted to 85 volumes (of Method 3 above) with over sixteen hundred items, and was once not inaccurately described as making the possessor independent of the zoological library at the local University. It was followed by a Second Series (Method 4) which at present holds over two hundred papers in eleven folders.

These considerations make it clear that arachnology does not confine its attractions to pleasant afternoons in woods or ditches, followed by even pleasanter hours with the microscope: it has aspects that can usefully occupy all hours and all seasons.

11

Arachnology in the Library

There have always been naturalists who have laid so great an emphasis on collecting and observing in the field that they have tended to turn themselves, and others, away from reading about animals, and almost to look down upon those who spend time with books. In the days when ecology was beginning to spread beyond the ranks of its devotees, one authority wrote, 'The idea that truth can be found in books is still widely prevalent, and the present dearth of illiterate men constitutes a real barrier to progress.' Nevertheless there is to be found in books much truth that an arachnologist must know, and which can be learnt by reading them. An arachnologist of today would find it hard to realize the position in which his predecessors of sixty or seventy years ago found themselves. They were usually told that they should consult Blackwall's *Spiders of Great Britain and Ireland* and Pickard-Cambridge's *Spiders of Dorset*, but they were not always informed that these books were also almost unobtainable. They are even rarer now, but their practical value has fallen because there are now books enough to start the beginner on his way and also to help the more experienced biologist.

The simplest and most satisfactory way to start is to visit the nearest Public Library, and there to scan the pages of any comprehensive 'Natural History'. These multi-volume works are not always very recent publications, but they will set the beginner's feet on the right path and may well be followed by a reading of the appropriate articles in any of the good encyclopaedias.

Preliminary reading of any kind may help the biologist to avoid useless repetition. Even today one may come across descriptions of quite elaborate pieces of research, carried out with the help of most sophisticated apparatus, and leading to a conclusion of moderate significance, already long-recognized as established fact.

The following notes are intended to lead the reader to the books that he may be advised to read, and also to help him to choose which of them he ought to possess for immediate reference. They have been arranged into three main groups of advancing standard.

Elementary

Savory, T. H. Spiders. Ginn. 1971. One of a series of 'First Researcher' booklets, intended for beginners. Its chief features are the unusually attractive illustrations assembled by the publishers, and, in the text, the emphasis laid on simple experiment.

Savory, T. H. The Spiders and Allied Orders of the British Isles. Warne. 1935. One of the 'Wayside and Woodland Series', this is an attempt to describe some of the commoner species of British spiders, over 200 of which are mentioned. There are keys for distinguishing the genera of the different families, and the coloured plates help in recognizing 63 species. There are also half-tone plates and drawings in the text. The illustrations of the genitalia were made from microscope transparencies, and do not always resemble the normal outward form. There are added adequate accounts of the British harvestmen and false-scorpions, all the species then known being included, and there is a section on the sea-spiders or Pycnogonida. Out of print.

Snow, K. R. The Arachnids: an Introduction. Routledge. 1970. A short account of the class, inspired by the appreciation of Arachnida by various examining bodies. Consequently mites, ticks and spiders occupy most of the text and the other orders are more briefly referred to. Limulus is included. There are 47 clear line drawings.

Intermediate

Bristowe, W. S. The World of Spiders. Collins. 1958. One of the 'New Naturalist' series, this is by far the best introduction to the study of spiders for the British arachnologist. The first chapter tells of the spider's appearances in myth, legend and superstition, and this is followed by accounts of structure, classification and evolution. Thereafter the families are taken in turn, and their members are treated not as units in a classificatory system, but as living organisms, whose behaviour deserves the closest attention. In the practice of spider-watching the author is supreme, telling us with a rare precision how a spider moves in a variety of circumstances, a limb by limb description which is choreography in reverse. It is superbly illustrated, with 40 half-tone plates and 116 text-figures in the inimitable style of Arthur Smith. Above all, the writer's personal experience and vital activity shines through his prose with a compelling enthusiasm that convinces the reader that the study of spiders can be enormous fun.

Cloudsley-Thompson, J. L. Spiders, Scorpions, Centipedes and Mites. Pergamon. 1968. There could be no better book to read as a successor to Bristowe than this rather more academic account, the first to be written on Arachnida by a professor of zoology. The author devotes seven chapters to different orders, and in each he describes their general behaviour, their feeding habits and mating methods in the manner of an experienced scientist. If Bristowe allows his readers to enjoy splashing in the shallows, Cloudsley shows us the ideal way of leading them into slightly deeper waters. His references spread to arachnids in every part of the world and to the literature of a correspondingly wide range. At its first appearance in 1958 the book seemed to be boldly experimental in its attempt to interest the general naturalists in the biology of the Arachnida; its continuing life shows how successful that experiment has been.

Gertsch, W. J. American Spiders. Van Nostrand. 1949. This member of the New Illustrated Naturalists' series is the American counterpart of *The World of Spiders*, and is cast in much the same mould. There are separate chapters on life-cycles, spinning, courtship and evolution, after which the systematic side of arachnology is compressed into four chapters. If not as light-hearted as Bristowe's book, it takes a rather broader view, with some early references to orders other than that of spiders, and introduces the reader to much that he will not easily find elsewhere. The book illustrates throughout its author's wide experience and deep knowledge of every aspect of the study of spiders. The illustrations are outstanding, with 32 plates in colour and 32 in gravure.

Locket, G. H. and Millidge, A. F. British Spiders. Ray Society. Vol 1, 1951; Vol 2, 1953; Vol 3, 1974. No student of British spiders serious enough to wish to put a name to every specimen that he catches can afford to be without these volumes, the work of two supreme authorities on the British spider fauna. When Volume 1 appeared after ten years of preparation, it was the first full account of its subject since *The Spiders of Dorset*, eighty years previously, and the eagerness with which it was acquired by naturalists everywhere proved how acute was the need for just such a book. After a history of arachnology in Britain, contributed by Dr Bristow, the families and the 591 species known in 1953 are described with notes on their rarity and distribution. Volume 3 is a supplement to the first two. For every species there is a drawing of the female epigyne and the male palp, with the result that everyone has a chance to discover the name of any mature specimen that he has come across. This almost incomparable book will remain the standard work on British spiders for many a year, and must be reckoned to belong to the same category as Simon's *Arachnides de France* and Roewer's *Weberknechte der Erde*.

Savory, T. H. The Biology of Spiders. This was an early attempt to bring together all relevant accounts of all aspects of spider biology: structure, growth, behaviour, distribution, classification and evolution. Useful as a summary of much scattered literature, with a sense of history that indicates the development of the science. Poorly illustrated, but within its range unlike other books in the English language.

Yaginuma, T. Spiders of Japan in Colour. Osaka. 1960. Although few of our arachnologists are able to read Japanese, this book deserves to be mentioned because of its fine production, with magnificent coloured plates, showing 320 spiders in all. The names of these are printed in Roman type. In Europe there has been no book on spiders to compare with this since *Die Arachniden* of 1848.

Advanced

Bonnet, P. Bibliographia Araneorum. Toulouse. 1945–1962. Professor Bonnet has devoted a large part of his life to the production of an immense bibliography, which included every printed book and paper about spiders, from the earliest times to 1939. Some eight thousand references are included and are also analysed to show the whereabouts of all work on each topic of structure, behaviour and distribution. The incredible work, which occupies five large volumes, is much more than a bibliography. There is a discussion of problems of nomenclature and a collection of biographies of 124 arachnologists, with portraits of nearly all of them, and a survey of the many attempts to classify the order satisfactorily.

Very few zoologists would have had the devotion and the courage needed to compile so exhaustive a work, which would only have been possible if its author had had a firm belief in the value of the task. Of this there is no doubt: to every araneist, amateur or professional, these volumes are beyond price.

Forster, R. R. The Spiders of New Zealand. Dunedin. 1967–. A fine work, which will ultimately cover some 1500 species of New Zealand spiders, is at present in course of production, and when complete will stand comparison with any other fauna in any country. To British readers the First Part is likely to be the most interesting; it covers structure, both external and internal, and describes the appearance and habits of a typical member of each family. Parts 2 and 3 are systematic, and are of great value in the light they throw on the problems of classification and evolution of the antipodean fauna.

Grassé, P. P. (Editor). *Traité de Zoologie: Tome VI.* Paris. 1949. This is almost certainly the most useful and the most readily accessible of the great and comprehensive accounts of the Arachnida. The class is given over six hundred pages, and the account of each class has been entrusted to a well-known arachnologist, specially suited for its description. Every aspect of the science is included and the illustrations, which number over five hundred, are admirable. Spiders receive about 150 pages. Although now more than twenty years old, the volume is of permanent value, and any enthusiast who takes the trouble to buy a copy for himself will never regret it.

Kukenthal, W. and Krumbach, T. (Editors). *Handbuch der Zoologie:* Vol. 3, Part 2. Berlin, 1932–38. In this immense German textbook the Arachnida receive that thoroughness of treatment that has always been associated with German zoology. As in the French publication mentioned above, each order has its own specialist-author, and spiders receive 250 pages out of a total of 656. The illustrations are copious. The second world war interrupted production before the scorpions were described: they were added later.

Savory, T. H. Spiders, Men and Scorpions. London. 1961. This is the only historical account of our science, and covers its growth from the days of Aristotle to about 1950. All orders are included, except the Acari. There are portraits of 12 arachnologists and short biographies of many more incorporated in the text. Out of print.

Savory, T. H. Arachnida. London. 1964. An attempt, by some critics said to have succeeded, to establish the claim of arachnology to the status of an independent science. It first describes the nine sub-divisions that are mentioned at the beginning of this book. The second part consists of 15 chapters, one for each order, and the remainder of the book consists of essays on a variety of arachnoid topics.

Specialized

Evans, G. O. and Browning, E. Pseudoscorpions. Linnean Society. 1954. This is No. 10 of the Linnean Society's *Synopses of the British Fauna* (First Series). It describes the 26 British species, with adequate drawings of their diagnostic details, and gives keys to the genera and species. It is an authoritative and valuable booklet of 23 pages, and of special interest since it is the only account of the native fauna since a similar publication by H. W. Kew in 1911.

Roewer, C. F. Die Weberknechte der Erde. Jena. 1923. This is standard work on the harvestmen of the world, and is a bulky volume of 1116 pages, not particularly easy to use because the descriptions and diagnoses are printed in an elaborate system of abbreviations, which save space but need to be learnt. Nevertheless it retains its position unchallenged at the head of the literature of Opiliones, and has been supplemented by its author in a number of more recent papers under the title of *Weitere Weberknechte.*

Savory, T. H. The Spider's Web. London. 1952. Consequent upon the fact that the production and use of silk is the spider's chief characteristic, this book deals with the nature of silk, the use of the silk glands and the different types of web produced. Much of it deals with the orb-web.

Sankey, J. H. P. and Savory, T. H. The British Harvestmen. Linnean Society's Synopsis: 2nd Series. 1974. Describes the habits and behaviour of the twenty-three British species, with descriptions of each and references to the work at present directed to the acquisition of further knowledge.

Tilquin, A. La Toile Geometrique. Paris. 1942. The author of this monograph was a French psychologist who devoted six years to studying the webs spun by the species *Argiope bruenichi.* Many hundreds of webs were watched, photographed and measured in an attempt to understand the details of the processes by which they are produced. The theories that he formed and the conclusions to which he came make a fascinating account of intensive study, and, if all the questions that the orb-web raises were not answered, a valuable quantity of information was made available for future students.

Weygoldt, P. The Biology of Pseudoscorpions. Harvard. 1970. This book, translated from the German original, is one of the great pieces of arachnology of the present century. It touches every aspect of its subject, showing clearly how remarkable the false-scorpion is in many respects. Special emphasis falls on the court-ship and mating, a process which included the making and using of a spermatophore; to be subsequently followed by the extra-ordinary development of the young and the maternal care that they receive. A regrettably short section is devoted to the relation of the order to the other orders of the class. No more stimulating book on Arachnida has appeared this century.

Witt, P. N., Reed, C. F. and Peakall, D. B. A Spider's Web. New York. 1968. Continuing the study of the orb-web, these authors have used the familiar species *Araneus diadematus,* as their chief source of facts, and have made a full study of the character of its web. From their collected data they have persuaded a computer to spin a web for them, but the instrument has not produced a very impressive fly-catcher. Much useful work has been done on the effect of drugs on the form of webs spun after injection. The book, which has good photographic illustrations, is an admirable example of a modern arachnological monograph.

Few of us are likely to have all the books that we should wish to possess, but our difficulties can be greatly reduced by the libraries. The first source of help comes from that branch of the Public Library service which can obtain on loan for any registered borrower almost any book in existence. A visit to the local Public Library and a few minutes conversation with one of the staff there will reveal to the serious arachnologist the surprising facilities that are to be had, merely for the asking. Sometimes the borrowed volume is so precious that it may not be carried home, but must be in the Library itself, but this small limitation is only a proof of the value of an organization that can make such rare works accessible to enthusiastic readers.

Nearly all learned societies possess specialized libraries, and the libraries of the Zoological Society of London and of the Linnean Society provide valuable privileges of fellowship of these bodies. Possibly it is true that the Zoological Society's library is richer in long runs of zoological journals, while the Linnean collection has a greater proportion of printed books; but this fact, if fact it be, is but a reason for application for fellowship of both Societies.

The British Arachnological Society has a large collection of re-printed papers, which members are able to borrow under ap-propriate conditions.

The conclusion is therefore obvious that joining at least one of these institutes, and ideally all three, is almost an essential for any arachnologist who has ambitions to pass beyond the elementary stage of the mildly interested spectator.

The existence of the following works may be drawn to the attention of any readers who are anxious to acquire further acquaintance with the literature of arachnology. Many of them are accessible in scientific libraries, and some may be borrowed through the Public Libraries service.

Baerg W J *The Tarantula* Kansas 1958

Berland L *Les Arachnides* Paris 1933

Berland L *Les Araignées* Paris 1945

Berland L *Les Scorpions* Paris 1948

Bristowe W S *The Comity of Spiders* Ray Society 1939/1942

Chamberlin J C *The Arachnid Order Chelonethida* Stanford 1931

Comstock J H *The Spider Book* New York 1948

McKeown K C *Australian Spiders* Sydney 1951

Nielsen E. *The Biology of Spiders* Copenhagen 1932

Savory T H *Instinctive Living* Pergamon 1959

Simon E. *Les Arachnides de France* Paris 1874–1937

Simon E *Histoire Naturelle des Araignées* Paris 1892–1903.

Thorpe R W and Wooderson W D *The Black Widow* Carolina 1945

Vachon M *Etudes sur les Scorpions* Algiers 1952

Vellard J *Le Venin des Araignées* Paris 1936

Warburton C *Arachnida* in *Cambridge Natural History* London 1909

12

Great Names of the Past

It is as a result of the unceasing labours, often under the most difficult of circumstances, of zoologists throughout the world that arachnology has become as well established and as clearly defined a science as entomology or ornithology and, in consequence, we, their successors, owe them a great debt. Interest therefore attaches to their lives, and a short account of those most eminent in their day is here contributed. Names of the living are not included.

C. A. Clerck (1709–1765) was a Swedish civil servant who had become interested in spiders after attending the lectures of Linnaeus. His work, *Svenska Spindlar*, 1757, became famous throughout Europe and about fifty spiders today bear the names that he was the first to give them.

Jean B. Lamarck (1744–1829) professor at the Museum of Natural History in Paris, founded the science of arachnology when he created from the Linnaean apterous insects his "Classe Troisième– Les Arachnides'. Though he did little more than suggest the existence of nineteen kinds of arachnids, he opened the way for real arachnologists such as Latreille and Walckenaer.

Pierre A. Latreille (1762–1833) was a priest who was imprisoned during the Revolution. In 1820 he became the first zoologist ever to hold a chair of entomology. He was above all a systematist, and produced several extensive works on classification. In these the classification of Arachnida made substantial progress.

Charles A. Walckenaer (1771–1852) was the first to separate a genus from the huge single genus Aranea of Linnaeus. In 1802 he wrote a work on the Fauna of Paris, naming 131 Parisian spiders. He retained all in one genus, but noted several groups in it. On these the genera of Latreille were largely founded.

John Blackwall (1790–1881), born in Manchester, was at first an ornithologist, who became interested in spiders about 1820. Retired from business, he lived at Llanwrst, near Bettws-y-coed, and became the real founder of British araneology. His great work, *The Spiders of Great Britain and Ireland* was published in 1861–63, and with its full descriptions and wonderful illustrations was for many years the standard work on the subject.

Nicholas M. Hentz (1797–1856) was born at Versailles and went to America in 1816, where he taught in several schools. Between 1820 and 1850 he wrote a series of papers describing and figuring American spiders. These, with a supplement, were collected and published in 1868. Hentz was one of the founders of American arachnology.

Anton Menge (1808–1880) was a graduate of Bonn, who taught in Danzig for over forty years. He is known for his large book, *Preussische Spinnen* of 1866–79, which included a history of arachnology, one of the first essays of the kind.

Octavius Pickard-Cambridge (1828–1917) was born at Bloxworth, Dorset, was ordained in 1858, and lived 'the uneventful life of a country parson' at Bloxworth for over fifty years. But if uneventful, his life was one of continuing service to his parish and his county, where he was at all times devoted to natural history. His interest in spiders dates from about 1850: he wrote over one hundred and sixty papers on arachnids, as well as his *Spiders of Dorset*, a successor to Blackwall's work, mentioned above. His ability, his personality and his achievements mark him out as one of the outstanding contributors to the growth and strength of arachnology.

T. T. Theodor Thorell (1830–1901) was lecturer and professor at Uppsala University. He was a man of wide learning, with a command of several languages. As a zoologist, he worked on marine organisms as well as arachnids, and one of his services to the science was his disentangling of the synonyms of European species.

He lived part of his life in Genoa, and wrote a series of papers on the collections in the Museum of that town.

Henry C. McCook (1837–1911) was an American clergyman and a naturalist chiefly interested in ants and spiders. He wrote nearly forty papers on the latter, and wrote, illustrated and himself published a three-volume work, *American Spiders and their Spinning-work*, 1889–94. This venture, of a kind that few naturalists are called upon to undertake, was a treatise on the nature and habits of spiders, far surpassing at the time anything else available in the English language.

George W. Peckham (1845–1914) was an American teacher and inspector of education. With his wife Elizabeth he provided the first instance of a collaboration between husband and wife in arachnology. They married in 1880, and between 1883 and 1909 over twenty papers bore their names as joint authors. They worked almost entirely on the family Salticidae, the courting dances of which they described and discussed in writings that have been quoted all over the world. They also investigated the memory and mental powers of spiders.

John H. Emerton (1847–1931), the only one of the early American arachnologists to visit Europe, devoted over sixty years to describing the spiders of America and various other regions to which he travelled. He was one of the first to study arctic spiders, and he wrote *Common Spiders of the United States*, the first popular book on the subject. All his work was accurate and reliable, and deserved its high reputation.

Eugene Simon (1848-1924) produced the first edition of the *Histoire Naturelle des Araignées* when he was only sixteen. This was his first step in his two great objectives, a systematic account of the spiders of the world and a description of the arachnids of France. All his life he devoted himself to these ends. He travelled widely, collecting spiders wherever he went, and conducting a correspondence with every arachnologist in the world. The first volume of *Les Arachnides de France* appeared in 1874; its sixth and last was completed after his death, in 1937. The publication of the second edition of the *Histoire Naturelle des Araignées* began in 1892 and was completed in 1903. Simon was quickly recognized as the world's supreme authority on spiders. His collection was contained in over twenty thousand tubes, the achievement of a man whose name and genius will never be forgotten.

Cecil Warburton (1854–1958) worked as a lecturer in Cambridge for the whole of his active life. He was the first to dissect the silk glands of spiders and the first to describe the cocoon-making of Agelena. He wrote the two hundred and fifty pages on Arachnida in the *Cambridge Natural History*, a contribution which for many years was the most authoritative source of information for British students. After the First World War he devoted himself to mites. He was also zoologist to the Agricultural Society, a post which he resigned at the age of 92. When he died at Grantchester at 104 he was probably the oldest man in Britain.

Friedrich Dahl (1856–1929) was a professor at Kiel for twenty years until in 1898 he became director of the laboratory at the Berlin Zoological Museum. He travelled widely and wrote many papers on various topics concerning spiders; notably he was one of the first to appreciate the ecological point of view. He wrote the account of the Lycosidae in *Die Tierwelt Deutschlands* (1927), in which publication other families were later described by his widow, Maria Dahl.

William J. Rainbow (1856–1919) was born in Yorkshire but lived most of his life in New Zealand and Australia. In 1895 he became entomologist to the Australian Museum in Sydney. He was virtually the founder of arachnology in the antipodes, being the author of over seventy publications, as well as a great Census of Australian Araneida, published in 1911.

Reginald I. Pocock (1863–1947), an Oxford zoologist who spent much of his life in the service of the British Museum at South Kensington. He made great contributions to our knowledge of several orders of Arachnida, especially the Theraphosid spiders, and also wrote on geographical distribution and the Carboniferous fossil species. He also contributed to the *Fauna of British India*.

H. Wallis Kew (1868–1930) was a naturalist of many interests, wholly concerned with Lincolnshire. For many years he was the chief authority on British false-scorpions: he was the first to describe their courtship and the first to distinguish their different instars, and was, indeed, the pioneer of precise knowledge of these arachnids.

Alexander I. Petrunkevitch (1875–1964) was born in the Ukraine, went to Freiburg University in 1899 and left for America in 1903. Here he filled various zoological posts until appointed professor at Yale in 1917. At Yale he gained the reputation of an inspiring teacher and the possessor of a penetrating intellect, devoted largely to the problems of arachnology. He was familiar with ten European languages, was blessed with unusual manual dexterity and endowed with the ability to continue working far into the night. In consequence he became recognized as the first arachnologist of the world. His many publications included a full account of the spiders of Puerto Rico, and several monographs on extinct and fossil arachnids. His classification of spiders was one of his outstanding works. In all this he attained the highest possible standards, such as is seldom to be found over so wide a range or to attract such universal admiration.

Ulrich Gerhardt (1875–1950), lecturer at Breslau and later director of anatomy at Halle, worked on several animal groups, specializing in spiders from 1921 to 1933. He concentrated his attention upon all aspects of the reproductive process with an exhaustive thoroughness over a wide range of families, in a manner that has never been equalled.

A. Randall Jackson (1877–1944), a medical practitioner of Chester, devoted all his leisure to Arachnida, and followed Pickard-Cambridge as the leading British authority. In particular the small spiders of the family Linyphiidae found in him an investigator with ideal qualities, and his studies of British and Arctic members of this group were of first importance. Apart from this, his outstanding service to arachnology was the ungrudging help he gave to everyone who sought his advice. So generous was he that his professional duties and his correspondence gave him no time to write the book on British spiders that all wished him to undertake. Like Simon, Jackson was a man of wide interests and precise knowledge, a scientist whose name will long be remembered.

Carl F. Roewer (1881–1963) graduated at Jena and later went to Hamburg, where in the Zoological Museum he was directed towards Arachnida by Karl Kraepelin. He worked at the classification of harvestmen until in 1923 he produced a thousand-page volume, *Die Weberknechte der Erde*. In 1933 he was appointed director of the Bremen Museum, and here he wrote great systematic works on Solifugae, Palpigradi and Pseudoscorpions. His last publication was the first volume of a world catalogue of spiders, *Katalog der Araneen*. He was a zoologist who realized that the task of compiling and co-ordinating is as important as original research, and that such constructive authorship deserves the gratitude of other workers.

Louis Fage (1883–1964) started as a marine zoologist at Banyuls, but moved to Paris and became a director of the Museum d'Histoire Naturelle. His zoological interests were wide, and among spiders his chief work was on spiders in caves and other special environments.

Lucien Berland (1888–1962) was a director of the Museum d'Histoire Naturelle in Paris, where he specialized in entomology. Like Fage, he was influenced by Simon, so that he also worked at arachnids, and concerned himself chiefly with the arachnid fauna of oceanic islands. Here he made important contributions to the zoo-geography of islands. His book, *Les Arachnides*, discussed the various phenomena, such as mimicry and courtship as found in the class.

Joseph C. Chamberlin (1898–1962) graduated at Stanford University, after leaving the American army. He occupied a number of academic and industrial posts, but as an undergraduate he had discovered how little was known about false-scorpions. His devotion to this order he retained throughout his life, and summarized his conclusions in a monograph, *The Arachnid Order Chelonethi*, 1931. He was one of the chief founders of modern chelonetology.

Glossary

Abdomen: the hinder part of an arachnid's body (not a stomach).

Appendage: general term for a leg or other limb attached to a body segment of an arthropod.

Autotomy: separation and sacrifice of an appendage.

Calamistrum: the comb of short spines on the metatarsus of the fourth legs of certain spiders.

Carapace: the protective upper surface of the prosoma, also called the shield.

Cephalothorax: the fore-part of an arachnid; the head and thorax together, also known as the prosoma.

Chelate: pincer-like, as the claws of a lobster.

Chelicerae: the jaws of an arachnid, weapons of attack and defence, often with venom glands within.

Chitin: a nitrogenous polysaccharide, the basis of the exoskeleton, and often hardened with calcium compounds.

Clypeus: the narrow strip of the carapace between the eyes and the bases of the chelicerae.

Colulus: a small conical protuberance, between the posterior spinnerets of spiders.

Coxa: the first segment of an arachnid's leg or palp, joining it to the body.

Cribellum: a sieve-like plate in front of the spinnerets of certain spiders.

Cuculus: the moveable 'hood' attached to the fore-edge of the carapace in the Ricinulei.

Distal: the portion of an appendage farthest from the middle line.

Diverticulum: a blindly ending, branching passage.

Ecdysis: casting of the outer skin or exoskeleton in growth, 'moulting'.

Epigyne: the external genitalia of a female arachnid.

Exoskeleton: the protecting and supporting structures, usually of chitin, outside the body.

Family: a group of related genera.

Femur: the third segment of an arachnid's leg or palp.

Flagellum: a whip-like addition to the end of the body. Also a group of setae on the chelicerae of false-scorpions and Solifugae.

Galea: spinneret on the movable finger of the chelicera in false-scorpions.

Genus: a group of related species.

Gnathobase: a projection from the coxa of a leg or palp, used in crushing food.

Instar: a stage in the development of an arthropod.

Labium: the lower lip or lower surface of the mouth, between the gnathobases or maxillae of the palpi or anterior legs.

Mandibles: in arachnids, an inaccurate name for the chelicerae.

Maxillae: in arachnids the alternative name for gnathobase, q.v.

Metatarsus: penultimate segment of an arachnid's leg.

Myrmecophilous: living unharmed among ants.

Ocelli: eyes of simple form, with smooth surface.

Operculum: chitinous covering to the genital orifice.

Palp, Palpus: leg-like organ in front of the first pair of legs, mainly tactile in function and often bearing the gnathobases.

Patella: short leg segment between the femur and tibia.

Pedicel (Pedicle): narrow 'waist', uniting prosoma and opisthosoma.

Pedipalp: a more precise name for palp or palpus.

Phoresy: clinging to another animal as a passenger, not as a parasite.

Phylum: the biggest division in classifying the animal kingdom.

Scopula: brush of setae at the end of a tarsus.

Serrula: a row of chitinous teeth, as on the chelicerae.

Seta: a hair-like sensory organ on the body and limbs of arachnids.

Somite: a segment of the body.

Spine: a stout seta.

Spiracle: a respiratory aperture, the origin of a tracheal tube.

Spermatheca: storage organ for the spermatozoa after their reception from the male.

Spermatophore: pillar-like secretion from some male arachnids, carrying a supply of spermatozoa.

Sternum: chitinous plate on lower surface of prosoma, lying between the coxae of the legs.

Stridulation: production of a sound by rubbing together two organs suitably surfaced for the work.

Tarsus: last segment of a leg or pedipalp.

Tergite: chitinous protective plate, dorsally situated and corresponding to a sternite below.

Tibia: the middle segment of an arachnid's leg or pedipalp.

Trachea: a tube carrying air into the body.

Trichobothrium: the very delicate setae believed to be able to respond to sound waves and so to possess an auditory function.

Trochanter: the second, ring-like, segment of an arachnid's leg or palp.

Index

Index

Index